Agriculture in Germany

Edited by
Stefan Tangermann

VERLAG

Die Deutsche Bibliothek - CIP-Einheitsaufnahme

Agriculture in Germany / ed. by Stefan Tangermann. - Frankfurt am Main: DLG-Verl., 2000
ISBN 3-7690-0588-0

Die Vervielfältigung und Übertragung einzelner Textabschnitte, Zeichnungen oder Bilder, auch für den Zweck der Unterrichtsgestaltung, gestattet das Urheberrecht nur, wenn sie mit dem Verlag vorher vereinbart wurden. Im Einzelfall muß über die Zahlung einer Gebühr für die Nutzung fremden geistigen Eigentums entschieden werden. Das gilt für die Vervielfältigung durch alle Verfahren, scannen der Abbildungen, einschließlich Speicherung, Veränderung, Manipulation im Computer und jede Übertragung auf Papier, Transparente, Filme, Bänder, Platten und andere Medien.

Alle Informationen und Hinweise ohne Gewähr und Haftung.

© 2000 DLG-Verlags-GmbH,
Eschborner Landstraße 122
60489 Frankfurt am Main
Telefon (069) 2 47 88-0
Telefax (069) 24 78 84 80
E-mail: dlg-verlag@dlg-frankfurt.de
Internet: http://www.dlg-verlag.de

Gedruckt auf chlorfrei gebleichtem Papier

Lektorat/Erfassung:
Umschlaggestaltung: Ralph Stegmaier, Offenbach am Main
Gesamtherstellung: Konrad Triltsch GmbH, 97199 Ochsenfurt-Hohestadt
Printed in Germany: ISBN 3-7690-0588-0

Preface

It is a good tradition in the International Association of Agricultural Economists that the country hosting the Conference provides delegates with information on its agriculture. We are delighted to continue this tradition in the context of the XXIVth International Conference of Agricultural Economists, Berlin, August 2000. In addition to the field trips during the 'German day' of the Conference, we hope that this book, and the Conference presentations based on it, serve this purpose well.

Like everywhere else in the world, the specific shape, structure and functioning of Germany's agricultural sector, of the upstream and downstream industries serving it, of the institutions around Germany's farming industry, and of agricultural policy in the country, are very much determined by history. All this is described well in this book. One most important and very specific event in Germany's recent history, i.e. re-unification of the nation in 1990, has created conditions that are found in no other part of the world. The New Laender in the East and the Old Laender in the West are still in the process of growing together. In agriculture, the structure that has emerged from the socialist regime in the East is rather different from that which has evolved over centuries in the West. Macro-economic conditions in the New Laender still diverge from those in the West of the country. The breathtaking moment of re-unification is well remembered by everybody in Germany, and the process of adjustments on both sides of what was the iron fence, running through the middle of the country, will take a long time. We want the participants in the Berlin Conference of the IAAE to get some feel for this process. Re-unification of Germany is therefore well reflected in the chapters of this book, in some cases explicitly, in others implicitly.

Among the many other characteristic features of Germany's agricultural sector, also covered in this book, is the institutional framework around it, including co-operatives and Chambers of Agriculture. A set of issues that may be of particular interest for visitors from outside Europe relates to the functioning of national agricultural policies in the framework of a Common Agricultural Policy pursued jointly by the fifteen Member States of the European Union. This book, therefore, provides pertinent information on and analysis of the complexities resulting from that policy constellation. Among the national policies still in the hands of the individual Member

States, rural development and social security are particularly relevant, and hence treated in the respective chapters of the book.

Germany has made, and continues to make, significant contributions to world-wide research and technology development in agriculture. Environmental issues receive growing attention in Germany's agriculture, and related policies have become a momentous element of the overall set of agricultural policies in the country. The position of German agriculture is increasingly determined by what happens in other elements of the value-added chain, starting from the provision of inputs into agriculture, and ending at the consumer's table. All these issues are discussed in the present book.

We are extremely grateful to our colleagues who have agreed to draft chapters for this book, in a very co-operative manner. The Federal Ministry of Food, Agriculture and Forestry has not only contributed chapters to the book, but also provided generous financial support, which is very much appreciated. The German Association of Agricultural Economists (Gesellschaft für Wirtschafts- und Sozialwissenschaften des Landbaues) has helped to organize this activity, and also contributed financially, based on a grant received from the Landwirtschaftliche Rentenbank, for which we are grateful. We are deeply indebted to Ann Hartell who has done an excellent job in editing the whole manuscript. Finally our thanks go to Petra Geile who has prepared the manuscript for publication. Without all this assistance it would have been impossible to produce this book.

We hope that the delegates of the XXIVth International Conference of Agricultural Economists at Berlin will remember the Conference well, and take home with them a good impression of German agriculture.

Stefan Tangermann (chairman of the editorial committee)
Institute of Agricultural Economics, University of Göttingen

Friedrich Kuhlmann
Institute of Farm Management, University of Giessen

P. Michael Schmitz
Institute of Agricultural Policy and Market Research, University of Giessen

Theodor Seegers
Federal Ministry of Food, Agriculture and Forestry, Bonn

Table of Contents

Preface			iii
Chapter	1	The History of German Agriculture *Werner Rösener*	1
Chapter	2	Agriculture in Germany – Production and Structure *Ursula Huber*	17
Chapter	3	German Agricultural Policy – Issues and Challenges *Volker Appel*	39
Chapter	4	Transformation of Agriculture in East Germany *Bernhard Forstner and Folkhard Isermeyer*	61
Chapter	5	Macroeconomic Framework and Implications for German Agriculture after Reunification *Dieter Kirschke and Steffen Noleppa*	91
Chapter	6	Rural Development in German – Issues and Policies *Eckhart Neander and Helmut Schrader*	111
Chapter	7	Social Policies for German Agriculture *Konrad Hagedorn and Peter Mehl*	135
Chapter	8	Institutional Characteristics of German Agriculture *Werner Großkopf*	167

Chapter	9	The Hierarchy of Agricultural Policies: European Union, Federal Republic and Federal States *Carsten Thoroe*	187
Chapter	10	The Role of Germany in the Common Agricultural Policy *Ulrich Koester*	209
Chapter	11	Agriculture and the Environment *Rainer Marggraf*	231
Chapter	12	Research and Technology in German Agriculture *Joachim von Braun and Matin Qaim*	255
Chapter	13	The Value-added Chain in the German Food Sector *Hannes Weindlmaier*	283

Chapter 1
The History of German Agriculture

Werner Rösener
Institute of History, University of Giessen

The Upswing in the High Middle Ages

Essential foundations for German agriculture were, without doubt, laid down the High Middle Ages. The structural changes in agriculture which took place from the 11th to the 13th centuries continued to be felt much later. What were these changes? Which economic and social changes then took place in agriculture? In the High Middle Ages, Germany and its neighbouring countries experienced a grand epoch of expansion and development in their economies, societies, and cultures. The enormous increase in population, advances in agriculture, and the rise of commerce and trade created foundations for the development of the character of European towns and the looming division of labour between the towns and the countryside. The upswing in the High Middle Ages was characterised by a marked growth in the population, an extensive increase in agricultural land, and a long-term increase in cereal prices; the first two changes were the most conspicuous.

From the 11th to the 14th centuries, the population of Germany grew threefold from four to twelve million, according to recent estimates. The population growth had its greatest expression in the expansion of cultivated land, particularly land under cereal cultivation. Consequently, the High Middle Ages became an era of intensive cultivation, during which the appearance of the cultivated land in Germany was completely changed, gaining an appearance which, in many areas, it basically retained up to the 19th century. The names of towns and configurations of fields still bear witness that they were settled during the High Middle Ages. The extensive development of land in the colonised area east of the Elbe river 'the German east settlement', contributed to the expansion of arable area in the old German settlement areas. In the old German landscapes, population pressure accelerated the pace of internal expansion of local boundaries, so that single farmsteads developed into hamlets and hamlets became villages. In the northern German marshlands, new settlement areas were established by drainage and the regulation of water by farmers hungry for land; in the low mountain ranges, such as the Black Forest and the Harz mountains, new hamlets and villages were systematically established.

The German eastern settlement was exceptionally complex, in terms of its course and outcome, and extended over a long period of time. Starting in the 10th century, the marches along the Elbe and Saale were extended eastward step by step, an important prerequisite for the German eastern settlement. True rural colonisation first began, however, in the 12th century, when the princes, the nobility, and the ecclesiastical manor lords made systematic efforts to secure farmers wanting to live in settlements and supported such farmers' establishment by means of advantageous settlement rights. Through the expansion of agricultural land, they intended to consolidate and extend their own territories. With national factors playing no role, Slav princes sought experienced farmers from the old settlement areas of Germany and The Netherlands willing to emigrate. Therefore, many settlers came from Flanders, Holland, and Westphalia to settle in the areas between the central Elbe region and the Erz mountains and areas further to the east. Using techniques for draining swamps and clearing forests, they laid down the long-term prerequisites for productive agriculture in the new settlements. In the contracts between the princes and the manor lords and the new settlers, one finds, first and foremost, a businessman who undertook the recruitment of settlers, distributed the available land, and supervised the setting up of farms. Frequently, he was entrusted with a large tract of land and granted some special rights, such as the exercise of local jurisdiction.

As incentive, favourable property rights, a great measure of freedom, fewer feudal taxes, and the abolition of arduous corvée were granted to the settlement-based farmers. The new settlements often granted free-farmer rights, which had developed in the villages on the cleared land of the old German areas. The large proportion of Flemish and Dutch farmers in the German eastern settlement groups, especially in Brandenburg and Saxon, resulted in the new settlers' rights being largely determined by these groups. The indigenous population was likewise affected by the favourable rights of the new settlers, as many Slav villagers were granted rights under 'German law' (ius theutonicum). On the one hand, significant improvement in property rights was associated with this status, and on the other, the village economic system was reorganised into the three-field system (i.e. Dreifelderwirtschaft). German eastern settlement and land utilisation were largely oriented to traditional forms of settlement. The economic farming unit, the Franconian hide of land approximately 25 ha in size, which became customary for these new settlements, was found from Upper Saxony and Silesia far into Poland. The most common types of rural settlement were typical villages, roadside villages, and woodland hide villages (collectively, called villages of colonisation).

The population increase and the expansion of settlements in the High Middle Ages were promoted by advances in the rural economy and in agricultural technology. Farms gradually became equipped with effective implements (ploughs, harrows, wagons, etc.) and increased their use of water power, as a result of the development of mills. Improved harnesses for horses and cattle made it possible to till the soil more intensively, and this resulted in higher yields. Yield increases also came in the 12th and 13th centuries when the three-field system replaced older systems, such as the grass field system and the two-field system. Greater yields, achieved as a result of the spread of the three-field system, also allowed a reduction in the proportion of cereal fields, to the benefit of vegetable and fruit cultivation and, in particular, of wine growing. The fundamental changes in the economy of the High Middle Ages were caused, above all, by the fact that, as a result of the spread of commerce and trade and advances in agriculture, the towns flourished and succeeded in implementing a monetary economy and exchange system. With the development of the town as the centre for handicraft production and the exchange of goods, a division of labour gradually developed between town and countryside, mediated by the market, in which the countryside supplied the town populations with food and received commercial goods in return. This division of labour quickly reached its limits where the agricultural economy had developed weak ties to markets.

The Agricultural Depression in the Late Middle Ages

Following the upswing in German agriculture in the High Middle Ages, in the 14th and 15th centuries there was a period marked by upheaval, characterised by Wilhelm Abel as the 'Late Middle Ages agricultural depression'. Clearly associated with this agricultural depression are the numerous wastelands that appeared during the late Middle Ages in many regions of Germany and which shaped the appearance of the settlements. Villages vanished into thin air, leaving only ruins in forests and fields. In earlier times, settlements were abandoned, but the number of the wastelands became much more noticeable in the Late Middle Ages. To comprehend the extent of these abandoned villages, one is dependent upon investigations into developments of settlements in different regions. Comparing the number of abandoned villages of the Late Middle Ages to the total number of settlements yields a ratio referred to as the 'wasteland quotient'. Calculating the wasteland quotient for Germany shows that about 25%, or every fourth settlement that existed during the expansion of land at the end of the High Middle Ages, was lost by the close of the Middle Ages and became wasteland. The frequency of wasteland differed markedly by individual region. The number of villages that were abandoned was particularly large in the Brandenburg March, in Hessen, and in south Lower Saxony, numbering more than 40% of existing settlements. In contrast, the number of wastelands was low in the northwestern German scattered settlement area and in the lower Rhine. In some cases when a village perished, its surrounding fields were taken under cultivation by neighbouring villages. When villages died out and their fields were abandoned, large tracts of land were left uncultivated. Thus the wasteland quotient is an indicator of the abandonment of agricultural land as well as of the villages themselves.

What were the causes of this amazing process of wasting which lastingly shaped the settlements in the German landscape? In the search for causes, a wealth of interesting explanations have been developed, which have allowed for much speculation. The most convincing explanation for the complex processes of wasteland formation in the Late Middle Ages is still given by the 'agricultural crisis theory' advocated by Wilhelm Abel and other agricultural historians. Despite a few objections and corrections which have been made to the theory, one can still adhere to its basic propositions. It considers the decline in population in the Late Middle Ages, the development of prices for agricultural and commercial products, and wages as central factors and places them in a reciprocal relationship.

The principal cause of wasteland creation in the Late Middle Ages was the decline in the population linked to the plague epidemic in the 14th century. With the first wave of the plague, between 1347 and 1351, approximately one-quarter of the population in many regions of middle Europe were victims of the dreadful epidemic. In the second half of the 14th century, there were additional waves, which further reduced the population, so that the total population of Germany declined by about one-third by 1450.

Wasteland was also a result of migration. Several factors were involved. The shift in the relationship between agricultural prices and the prices for commercial products reduced agricultural incomes and sparked migration from the countryside. Farmers left the settlements and set out for neighbouring towns which had lost many inhabitants to the plague, or they sought villages with better economic conditions. Wastelands were particularly more frequent in highland areas, on poorer soils, or in regions where settlements were established later. Obviously farmers in the Late Middle Ages migrated from poorer soils and problematic mountainous landscapes to more productive soils and/or areas with means of transport in the vicinity (e.g. valley basins).

Because of the loss of population and the resulting reduced demand for cereals, there was a long-lasting decline in cereal prices in the second half of the 14th century. As cereals were among those basic foods not subject to significant fluctuations in consumption, a surplus arose. The price of cereals sank, causing severe income losses for farmers and manor lords who produced cereals for the market. The reduced demand for basic foods and the low cereal prices also resulted in striking changes in land utilisation in the Late Middle Ages. Poor land was abandoned and only improved arable farming techniques were used, so that overall there was a rise in productivity. The reduction in arable land areas was accompanied by an expansion of pasture land; as a result the fodder base for keeping cattle was increased. Therefore keeping cattle spread in many regions in the Late Middle Ages at the expense of arable farming, and made an attractive supply of meat products available to the population. Moreover, the most important branch of livestock breeding was cattle, as it was easier to bring cattle through the winter than pigs. The reduction in cereal cultivation mirrors the expansion of the cultivation of other field crops, whether for food or industry. The cultivation of plants such as madder or willow supplied raw materials for the developing textile trade, while wine, hops, fruits, and vegetables extended the range of food for the table.

The Expansion of Land and Cyclical Movements in Agricultural Markets in the Early Modern Period

After a period of stagnation, Germany's population again substantially increased in the early 16th century. The increase in population was reflected, in addition to the growth of villages and towns, in the expansion of agricultural land and in the acquisition of new land. Cultivation was again carried out on ground which had previously been cultivated but was abandoned in the Late Middle Ages. Land reclamation frequently began near existing villages and expanded to neighbouring fields. The recultivated fields did not meet the needs of the growing population, particularly as some fields were now forested and protected by forestry regulations. The population increase forced development away from the reforested wastelands and into new land in the forest and mountainous regions of the German low mountain ranges, as well as in marsh areas near rivers and seas. Using the amount of cultivated land as an indicator, one can assume that around 1560 the German population had returned to pre-plague levels. Wherever possible, marshland and wasteland was cultivated, pasture was converted to arable land, and forests were cleared. Deforestation increased in the 16th and the 18th centuries to such an extent that in some territories, the princes were forced to issue prohibitions against clearing and to limit the number of cattle. Farmers were under enormous pressure and had to find other sources of food for themselves and their families.

On the German North Sea coast, marshland drainage and dyke building were aggressively pursued. Following substantial losses of land in the 14th and 15th century around the Heligoland Bight when many villages and fields were flooded, making dykes and draining land began again in the 16th century. In Jever, in the Harl basin, and at other sites along the coast, approximately 40,0000 ha of marshland were reclaimed in the 16th and 17th centuries, more than two-thirds of the amount of wasteland reclaimed in Lower Saxony from the 13th to the 19th centuries. Extensive areas were also reclaimed from the sea in neighbouring Schleswig-Holstein.

Inland, remarkable social differentiation between the old farmers (tenant farmers and land heirs) and the newly appearing social classes occurred in association with the expansion of cultivated area and the arrival of new settlers in northern Germany and the east German marches. These new arrivals were cotters and crofters and differed from other rural social groups both in the smallness of their enterprises and by their diminished rights. Their legal disadvantage was, above all, manifested in the restrictions on land use in the marches. The work of cultivation, done by both the old farmers as well as the subsequent settlers, was frequently carried out at the outskirts of the existing fields and then reached into common land areas.

Following a depression in the 17th century which was linked to the heavy losses during the Thirty Years War, in the 18th century a new upswing in the rural economy began, associated with a significant population increase. Between 1740 and 1805, for example, the population in the kingdom of Prussia more than doubled. The consequences of this increase were considerable increases in agricultural prices and general improvement in agriculture. The increase in agricultural production, forced by the growing population, could be achieved either by expansion of arable areas or improvement in the methods of production, that is via additional expenditure of work and capital. First and foremost, one should note the practice of planting parts of fallow fields and thereby achieving an improved three-field system which spread in the second half of the 18th century. Success in the implementation of this improved form of land utilisation was, however, frequently impeded by land constraints and established common land rights.

Manor Lords and Estate Owners

Up until the liberation of farmers in the 19th century, the state of agriculture in early modern Germany was characterised by the contrast between the manor lords and the estate owners. What elements were associated with these different sovereign and economic systems? In which areas of Germany were the manor lords and estate owners represented? The border between the two agricultural systems corresponded approximately with the river Elbe. Estate ownership prevailed from the Elbe eastward: from Holstein via Mecklenburg, Pommern, and East Prussia to Poland, and on via the Brandenburg March, Lusatia, Bohemia, and Moravia to Rumania. Regions west of the Elbe were commonly controlled by manor lords; however, land ownership took different forms in the individual provinces, so that zones with different agrarian systems, which could be more or less delineated, emerged in early modern Germany.

In the manor lord system, land rights were split between the superior ownership by the manor lords and the ownership of use by the farmers. Western Germany was characterised by small rural farms, which paid their manor lords mainly with money and payments in kind. A few fields in these enterprises belonged to the manor lords and were farmed by them, but these were generally not extensive areas. In the east, the large enterprises of the estate owners dominated, and farmers were obliged to render corvée along with their commodity obligations. In these large agricultural enterprises, large tracts of farm land stood at the centre of the estate owner's domain.

The position and significance of the rural communes were different under the two systems. East of the Elbe, rural communities had close ties to the estate owners; many communes were allocated to a single estate owner. At the same time, estate owners possessed police powers and lower judicial authority in their villages and farmers were considered inherited subjects. In the villages west of the Elbe, the sovereignty structure was generally highly fragmented; each village often had several manor lords with property and rights at their disposal. The villages in western and southern Germany had, as a rule, far greater responsibilities and freedoms than villages in the east.

This bipartite German agrarian structure greatly moulded not only the sovereignty relationship, but also the social and economic conditions of the farmers in the respective areas. In which period did this agrarian dualism develop? The system of estate ownership east of the Elbe certainly does not go back to the High Middle Ages, as the areas east of the Elbe were fully opened later. In the course of this development of the land,

German farmers in the east achieved a better economic, social, and legal position than that persisting in the old German settlement areas.

The 15th and 16th centuries constitute a decisive period, in which agrarian relationships developed quite differently in eastern and western Germany. In the course of the difficult economic processes at the end of the Middle Ages, characterised by a string of epidemics, increasing wastelands, and declining agricultural incomes, estate owners took different measures which, for many farmers east of the Elbe, meant starting on the path towards serfdom. Instead of removing feudal obligations, as was done in England, the estate owners demanded more work from the farmers and decreased their mobility, and extended demesnes at the cost of farmland. In the 16th century, freedom of movement in the eastern rural areas was increasingly restricted and farmers were bound to the soil. During the 18th century, estate ownership underwent its greatest development. The noble estate owners in Mecklenburg and Brandenburg were now relatively independent lords in their districts and, to some extent, used their power ruthlessly to achieve their objectives and expand the economic activity of their estates.

Rural Inheritance Customs

Rural inheritance customs were a formative influence on agrarian structure in the various regions of Germany. The size of agrarian families and the scale of farming enterprises were for many centuries dependent on the various inheritance rights and customs. As a rule, areas with mainly divisible inheritance tended towards the small farm system and the creation of small farm families, while areas with inheritance of intact property tended towards larger households. There have been abundant hypotheses advanced concerning the genesis of the regionally divergent inheritance customs in Germany, and indeed many have a certain core of truth, but none can claim absolute validity for all German agricultural landscapes. Without a doubt, the forms of agrarian inheritance owe their existence to the interaction of several factors. It is, however, impossible to project the origin of the regional distribution of inheritance customs as they existed in the 19th century in Germany to the High Middle Ages. That is to say, the regional distribution of particular inheritance customs was gradually established in the course of several centuries. At the beginning of the 19th century, the inheritance of intact property was most notable in northwest Germany, in east Germany, and in old Bavaria, while divisible inheritance dominated, above all, in Rhineland, Franconia, and southwestern Germany.

That intact inheritance derived from the old German law of the people, an idea popular in the Third Reich and which served as the ideological underpinning of the Reich's 1933 hereditary farm law, cannot be maintained academically. Evidence shows that the division of property was altogether possible throughout the German realm and was practised rather often, even in the Scandinavian countries. When analysing rural inheritance customs in Germany, one must take into account three factors: (1) the influence of manor lords, (2) the natural soil and climatic conditions and the economic situation, and (3) the rural mentality regarding inheritance.

The manor lords generally pressed for farmsteads to be taken by a single heir, whether the oldest or the youngest, in order to ensure efficiency of the farmsteads by maintaining suitably large enterprises. Under the influence of the manor lords, areas developed in Bavaria and Westphalian with intact property inheritance. Under conditions which supported wider distribution of property (dense population, good soil conditions, intensive cultivation), the manor lords were apparently unable to push their demand for the intact inheritance of farmsteads. In cases in which farmsteads were held by the lord and were not freely heritable, as a rule manor lords transferred the use of the property following the death of the current tenant to only one son. Frequently a successor was named before a farmer's death, and the old farmer retired after having signed a contract transferring his land to his successor. The principle of indivisibility of farmsteads was retained in many villages even after the influence of the manor lord ceased to exist and farmers could freely dispose of their farmsteads. In this way, in large parts of Westphalia and old Bavaria the transfer of intact property has been continued as the rural inheritance practice until today. In contrast, in areas such as Württemberg, where the principle of division of inheritance was retained, numerous small farm enterprises and small household families came into being. In regions where division of property predominated, large nucleated villages with acquired village land came into being, there was an increase in the population, a decline in agricultural enterprises, and growth of a small farmer class which was dependent on supplementary income.

Agricultural Reform and the Liberation of the Farmers

The liberation of the farmers was, without doubt, the most far-reaching event of modern German social history. Seldom have governmental measures had such a decisive effect on rural society and the economy as the agricultural reform and liberation of farmers that took place in Germany between 1790 and 1850 and paved the way for market-oriented agriculture. The multifarious measures which were implemented during this period that brought the abolition of the feudal agricultural order are referred to as 'the liberation of the farmers'. Contrariwise, in the language of the day, the liberation of farmers on estates was referred to as 'regulation', and that of farmers tied to manor lords as 'discharging' or 'release from the land'.

The 'liberation of the farmers' generally refers to three reform processes: discharge, division of communal property, and separation and coupling. Discharge means the release of farmers from diverse manorial dependencies, i.e. it referred to the liberation of farmers in the strictest sense. Division of common property meant the dividing up of the areas used communally in the vicinity of the village, also referred to as common land or march, which were characterised by the individuals' entitlement to their use. Finally, separation and coupling meant the integration of small plots of land that had been split from the communally cultivated village fields to create larger field complexes. These areas, freed from restrictions, could be independently cultivated by a single enterprise.

Although the dissolving of the manor lord-farmer relationship was at the centre of agricultural reform in Germany in the 19th century, measures were not limited to this. Generally, other reform processes were associated, such as the abolition of the joint economic systems of the farmers in arable fields and common land. The abolition of restrictions on fields and the binding regulations governing cultivation, together with the establishment of larger fields as a result of coupling made possible individual cultivation with improved methods. The splitting of the common areas and the ending of rural pasture cooperatives provided individual farm enterprises with new land, which they first ameliorated. Consequently, dependency on the manor lords and estate owners was eliminated, and the abolition of bonds with the cooperative brought further independence. The splitting of common property was a significant liberation for many farmers in the old villages; they were set free from collective dependency. Areas used cooperatively and areas of common pasturage sharply restricted the freedom of individual farmers, leaving them unable to change to new methods of cultivation. In some districts, the restrictions on fields were indeed more oppressive than the obligations placed on them by the old manorial system.

The liberation of German farmers was, incidentally, closely linked to the extensive reform work that modernised the state and society in the early 19th century, most evident in the Prussian reforms. Therefore, one must view the liberation of farmers as the first step in a modernisation process in the agricultural sphere that continues today. The liberation of farmers ultimately had as its objective the integration of the rural population, about 80% of the total German population in 1880, into the new political order of the 19th and 20th centuries.

The pace of the liberation of the German farmers was greatly influenced by events taking place in neighbouring France. During the reign of Napoleon, the German Confederation of the Rhine states was closely allied with France. This gave rise to the start of important agrarian reforms. In the kingdom of Prussia, the breakthrough came between 1807 and 1811, in conjunction with the Stein-Hardenberg reforms. The edict issued in 1807 for the liberation of farmers heralded the abolition of all tenant subservience; farmers received personal freedom and the right of freedom of movement. At the same time, all restrictions deriving from social standing were abolished, so that noblemen and farmers alike were able to enter an occupation independent of title. In 1808, Stein succeeded in carrying through a provision for East Prussian farmers that for the first time initiated a true liberation. The regulatory edict of 1811 then granted the restrictions were lifted if farmers transferred one-third or one-half of their land. This edict affected about half of the Prussian farmers who owned a horse-drawn carriage; the large number of hereditary tenant farmers were included later.

As previously mentioned, the agrarian system of the Confederation of the Rhine states was based on the manor lord system. The turning point for the abolition of the old agrarian system occurred much earlier here than in East Prussia. For example, in the Grand Duchy of Berg, in 1808 the ties of all farmers to their manor lord were radically abolished and replaced by redemption payments. Between the Vienna Congress (1815) and the revolutionary year of 1848, the liberation of the farmers in western and southern Germany accelerated, although with some delays. In Prussia, as well, the disencumbering of land proceeded, but this was accompanied by a substantial loss of land for the farmers. The revolutionary year of 1848 heralded the last phase of farmer liberation: in the spring the parliaments of the states in southern Germany resolved the abolition of those feudal burdens that still existed. In the same year, the Frankfurt National Assembly drew up a catalogue of basic rights, in which all relationships of subservience and bondage were abolished. All fees and services encumbering the land could be removed; the patrimonial courts and the personal fees resulting from ties to the manor lords or estate owners were abolished without compensation. Although redemption payments continued for decades, true liberation of farmers had been achieved. A primary consequence of the agrarian reforms was that the rural population and the rural economy were integrated into the German state and society of the 19th and 20th centuries. Agriculture was therefore firmly incorporated in the market economy of the newly formed German empire which, in principle, eliminated obstacles to the possession of land.

Rural Economy in the Third Reich

The agrarian policy of the Third Reich (1933-1945) influenced the German rural economy in a much more powerful way than the agrarian policy measures of the Weimar Republic. Farmers' sense of self-worth had suffered a great deal as a result of industrialisation and the associated social and economic structural changes. Since the middle of the 19th century, agriculture had been losing significance in the overall national economy. Although agricultural production had greatly increased, the share of the agricultural sector in the national product had halved between 1850 and 1913 (from 46% to 23%). From 1924 to 1932, 80,000 ha of agricultural acreage in Germany had been sold at compulsory auction; 30,000 farmsteads, predominantly small- and medium-sized enterprises which were mostly inherited, family properties, were liquidated. The downward pressure on prices by the world market, declining domestic demand, and the burden of interest from the capital market, brought about a decline in rural incomes from 1929 to 1932 of almost 40%. The liquidation of enterprises affected the small- and medium-sized enterprises most of all, while governmental remedial measures mainly benefitted farm enterprises in the east. With their 'blood and earth' ideology, the National Socialist party (NS) offered farmers a supposed positive perspective for the future, promising state protection and a higher social standing for the rural population.

In 1933, farmers were by and large disappointed by the relative lack of success of those representing their interests. Rooted in a conservative way of thinking, they were largely receptive in terms of mentality and basic attitude to the NS slogans of political agitation. Even the racist and anti-semitic elements of the ideology of 'blood and soil' played a significant role in this connection; the hatred of the 'animal Jews' was as equally disseminated amongst farmers as was the fear of socialism and bolshevism. The slogans of unity and corporative politics also brought the NS success with the rural population because they nostalgically evoked the community harmony of the preindustrial period when the agricultural sector was dominant. The agricultural policy arm of the Party (NSDAP) ostensibly stood up for the interests of farmers more energetically than other parties, and in their functionaries, presented a leadership cadre for the rural population that came from the ranks of farmers, particularly young farmers. The entire agricultural organisation was infiltrated, mobilised, and used for the instrumentalisation of the goals of the Third Reich and it was eventually brought completely into line with the Party.

Walther Darré became the most important representative and organiser of NSDAP policy in May 1993 when he was made the Reich's farmers' leader. In this position he was head of the federations and the farming cooperatives. On 29 June 1933, Darré also took over the Reich's Ministry for Food and Agriculture, and the 'farming body' became the 'farmers and agricultural workers'. With the law concerning the 'provisional organisation of the Reich's farmers and agricultural workers and measures for market and price regulation of agricultural products' enacted in September 1933, a powerful farmer's syndicate was created in Germany under the catchword 'self-administration', in which 17 million members were united. Not only farmers and agriculturists belonged, but also food producers and those employed in trade of agricultural products.

A key element of the NS agricultural policy was the Reich's farm inheritance law of 29 September 1933. All agricultural enterprises that offered at least subsistence (i.e. provided for the existence of a family), assumed to be about 7.5 ha in size, were declared hereditary farms. The upper area limit was set at 125 ha. Larger enterprises could, under special circumstances, be included. In order to ensure the continuity of ownership in individual families, the precise succession was laid down. The hereditary farms were subject to compulsory intact inheritance law, i.e. farmsteads went undivided to one of the legitimate claimants as principal heir. At the same time, male relatives had absolute priority ahead of any female offspring. Disposal, encumbrance, and leasing of hereditary farms was, except for specified exceptions, not allowed. Above all, any existing encumbrance on hereditary farms was to be repaid to avoid critical financial bottlenecks.

For NS agricultural policy from 1933 to 1939, national food security was the focus, to guarantee that in the event of a foreign policy conflict the German population could be fed from German soil. The achievement of this goal was embodied in the phrase 'the farmers' battle', which was proclaimed in propaganda. The German rural economy was organised so that general production targets were predetermined by the state and the Party. A large measure of decision-making freedom was given to the private sector, so that the individual farmer was able to decide independently whether he wished to use fertiliser and agricultural machinery. Production was controlled by general propaganda and individual counselling. One tried, above all, to emphasise the production of those fruits and products for which Germany had a substantial contribution from abroad (e.g. fats, meats, vegetables). Overall, the efforts at that time to increase agricultural production had only limited success. The improvement in agricultural production was, moreover, partly offset by a population increase, thus the level of German food self-sufficiency rose by only a few percentage points under the Reich. German agriculture from 1939 to 1945 was subject to the special conditions of World War II, and therefore will not be considered here.

References

Wilhelm Abel (1976). Die Wüstungen des ausgehenden Mittelalters, Stuttgart [3].
Wilhelm Abel (1978). Geschichte der deutschen Landwirtschaft vom frühen Mittelalter bis zum 19. Jahrhundert, Stuttgart [3].
Wilhelm Abel (1978). Agrarkrisen und Agrarkonjunktur, Hamburg u.a.[3].

Walter Achilles (1991). Landwirtschaft in der Frühen Neuzeit, München 1991.

Gustavo Corni / Horst Gies (1994). Blut und Boden. Rassenideologie und Agrarpolitik im Staat Hitlers, Idstein.

Christof Dipper (1980). Die Bauernbefreiung in Deutschland, 1790-1850, Stuttgart u.a.

Horst Gies (1979). Aufgaben und Probleme der nationalsozialistischen Ernährungswirtschaft 1933-1939, in: Vierteljahrsschrift für Sozial- und Wirtschaftsgeschichte 66, P. 466-499.

Friedrich-Wilheim Henning (1978/79). Landwirtschaft und ländliche Gesellschaft in Deutschland 1-2, Paderborn u.a. 1978/79.

Friedrich Lütge (1967). Geschichte der deutschen Agrarverfassung vom frühen Mittelalter bis zum 19. Jahrhundert, Stuttgart [2].

Helmut Röhm (1957). Die Vererbung des landwirtschaftlichen Grundeigentums in Baden-Württenberg, Remagen.

Werner Rösener (1985). Bauern im Mittelalter, Munich.

Werner Rösener (1991). Grundherrschaft im Wandel, Göttingen.

Werner Rösener (1992). Agrarwirtschaft, Agrarverfassung und ländliche Gesellschaft im Mittelalter, München.

Werner Rösener (1993). Die Bauern in der europäischen Geschichte, Munich.

Alois Seidel (1995). Deutsche Agrargeschichte, Freising 1995.

Chapter 2

Agriculture in Germany – Production and Structure

Ursula Huber
Federal Ministry of Food, Agriculture and Forestry, Bonn

Site Conditions

The territory of the Federal Republic of Germany comprises an area of around 357,000 square kilometres, 50% of which is used for agriculture and 30% for forestry. The site conditions for agriculture and forestry are marked by natural diversity. The production area stretches from the northern German lowlands with extensive moors and heaths, and the Mecklenburg lake district, across a landscape of low mountain ranges with altitudes of up to 1,500 metres and river valleys, some of them deeply carved, with fertile soils to the Alps, where agriculture (alpine farming) is practised up to altitudes of 2,000 metres. This diversity is also reflected in the soil quality. Fertile, loamy to clay soils are found at the southern edge of the northern German lowlands (Börden) and in some parts of southern Germany. Less fertile sandy and stony soils are found in the moraine area and in low and high mountain ranges.

Climatically, Germany lies in a temperate zone heavily influenced by the gulf stream and experiences frequent weather changes. Westerly winds and precipitation in all seasons are characteristic. In the northern German lowlands, annual precipitation is below 500 to 700 mm; in the lower mountain ranges from 700 to over 1,500 mm; and in the Alps, over 2,000 mm. The mean annual temperature is + 9°C. Average temperatures in January, the coldest month of the year, range from + 1.5°C to –0.5°C in the lowlands, and can drop below –6°C in the mountains depending on the altitude. In the northern German lowlands, July temperatures average + 17°C to + 18°C and in the upper Rhine valley up to + 20°C.

These natural conditions determine the spatial distribution of agricultural production centres. Cattle breeding and cattle farming focusing on dairy production prevails in the large grassland areas of the northern German coastal area, in the foothills of the Alps in southern Germany, and in the lower mountain ranges. Today, sheep farming is again practiced on the particularly rough grassland sites of the lower mountain ranges. A favourable wine climate prevails in the upper and middle part of the Rhine and in the lower reaches of its tributaries Neckar, Main, Nahe, Mosel, and Ahr as well as in the Saale-Unstrut area and in the upper course of the Elbe.

Suitable climatic preconditions can also be found for the regionally restricted cultivation of tobacco, hops, fruit, and vegetables. Tobacco is cultivated throughout Germany, but particularly in Rhineland-Palatinate and in Baden-Württemberg. The hop-growing region is concentrated in Hallertau, north of Munich, the Tettnang region (Lake Constance) and on the Dresden area. Due to market proximity, vegetable growing and horticulture is concentrated around the large conurbations, for example, south of Hamburg, in the Oderbruch, in the vicinity of Bonn, and in the triangle of Rhine, Main, and Neckar. Fruit cultivation areas are located in the Alte Land, west of Hamburg, the lower Rhine region, in the Havelland, south of Halle and Dresden, around Lake Constance, and in the Rhine-Neckar region.

Importance of Agriculture in Germany

Gauged by certain economic parametres, agriculture has clearly lost importance over time. Whereas the share of agriculture, including forestry and fisheries, in the gross value added still accounted for 5.8% in the old Laender in 1960, it only amounted to 2.2% in 1980 and still continues to decline. In 1991, the share of agriculture, including forestry and fisheries, in gross value added totalled around 1.4% nationwide and has since dropped to less than 1.3%. The number of persons employed in the sector has declined analogously to the sector's share. In addition, substantial job cuts occurred in the new Laender, especially shortly after unification. The share of persons engaged in agriculture, forestry, and fisheries in the overall activity rate in Germany has therefore decreased from 4.2% (1991) to 2.9% (1998). Further declines can be expected.

This view is confined to agriculture, and therefore does not include the up- and downstream economic sectors, so that the importance of the agricultural sector for the overall economy is underrated. If one includes the entire agribusiness sector, its share in the gross value added is roughly 6.5% (1997). The same applies to the importance of the agricultural sector for the labour market. Taking all agribusiness into account, 11.5% of the active population in 1997, or around 4.1 million persons, were engaged in the agribusiness sector. Thus, every ninth job in Germany is directly or indirectly related to agricultural production.

In view of the fact that 21% of the sales proceeds of German agriculture come from exports, foreign trade is of particular importance for the agricultural sector. In 1998 Germany ranked fourth in global agricultural trade with an export value of 47.3 billion DM, after the USA, France, and The Netherlands. As a result, around 6% of the agricultural and food commodities in world trade originate in Germany. In terms of value this was more than nine times the 1970 level and more than double that of 1980. Regarding agricultural imports, Germany ranks first worldwide with 77.6 billion DM. Deducting export value, Germany is the second largest net importer in agricultural trade worldwide after Japan.

The key export products are dairy products, whose export value amounted to over DM 7 billion in 1998, followed by meat and meat products with around DM 4.0 billion. Fruit and tropical fruits dominated imports with around 9 billion DM in value. Meat and meat products ranked second with DM 8.5 billion.

Seen in purely quantitative terms the role played by agriculture in Germany is underestimated. With 230 inhabitants per km^2, Germany is a densely populated country. At the same time 80% of the country is used for agriculture and forestry. This makes it absolutely necessary to reconcile the demands and requirements of a large population with the economic interests of farming and forestry.

Agricultural Production

Crop Production

Arable crops In 1998, the utilised agricultural area (UAA) accounted for 17.4 mio. ha in Germany. In the EU, Germany thus ranks third after France (30.2 mio. ha) and Spain (29.6 mio. ha), ahead of the United Kingdom (15.9 mio. ha) and Italy (15.1 mio. ha).

UAA in Germany is mostly used for arable crops (around 68%). Around 30% is used as grassland. The share of grassland is highest in the city states and in the Laender of Schleswig-Holstein and Baden-Württemberg, and lowest in Saxony-Anhalt. All in all, there are great differences between the new and old Laender in the shares of arable land and grassland in the UAA. Whereas in the new Laender the share of arable land in the UAA amounts to 79.7%, in the old Laender it is only 63.0%. Accordingly the share of grassland in the new Laender is only 19.9% compared with 35.3% in the old Laender. This pattern of a greater share of arable land in the UAA in the new Laender existed prior to German unification (see table 1).

Table 1:
Shares of arable land and grassland in the new and old Laender

	1970	1980	1985	1990	1995	1998
New Laender (and/or former GDR)						
Share of arable land in UAA	73.5	75.9	75.8	76.0	80.1	79.7
Share of grassland in UAA	23.4	19.7	20.1	20.2	19.5	19.9
Old Laender						
Share of arable land in UAA	55.5	59.4	60.3	61.2	62.7	63.0
Share of grassland in UAA	40.5	38.8	38.0	37.1	35.6	35.3

Source: Federal Statistical Office: Fachserie 3 Reihe 3.1.2: Utilised Agricultural Area, several annual series; own calculations.

The share of arable land in UAA increased in both parts of Germany over time, while the pro-rata share of grassland changed little in the new Laender, it markedly declined in the old Laender. Cereals cultivation dominated the arable land throughout Germany with 59.3%, with wheat being the principal cereal species. Commercial crops (rape and colza seeds, sunflower seed, flax, hops, etc.) and forage plants (silage maize, clover, grass-clover, alfafa) have shares of 10.3% and 14.9%, respectively. Rape is the key crop among the commercial crops, and silage maize among the forage plants. The share of root crops (especially sugar beet and potatoes) in arable land amounts to 6.9%, and that of pulses (especially field peas and field beans) to 1.9%.

In the old Laender, the share of cereals area in the entire arable land reached a maximum of 71.8% in 1979, then dropped over the course of time to 56%, and now slightly exceeds 60%. In the former GDR, by contrast, the share of cereals in the arable land was relatively low, around 50%, with the share of root crops relatively high, around 14%, primarily due to state-promoted potato cultivation. As the state requirements were abandoned with unification, the share of root crops have meanwhile fallen to 4.4%, the share of cereals rose to 56.6%, and the share of commercial crops to 14.8%.

Progress in production engineering and breeding led to marked yield increases over the past few decades. In the case of cereals, for example, yields in the old Laender increased from 28.7 dt/ha to around 54.7 dt/ha from 1950 to the late 1980s (an annual average increase of 2.4%). This growth rate has levelled off slightly to only 1.5% per year in the 1990s. Meanwhile, the average cereals yield in the old Laender is around 66.3 dt/ha (average of 1997-1998). The overall cereals yield level was far lower in the former GDR; around 1960 it totalled 24.5 dt/ha and rose by 2.1% per year on average to 43.5 dt/ha by the end of the 1980s. Since unification the cereals yield has risen by 3.4% annually to around 60.2 dt/ha thus approximating the level in the old Laender. This is due to the improved supply of fertilizers and plant protection products, the availability of better crop varieties, and better production engineering.

Rape ranks first in Germany among the oleaginous fruits. Especially in the 1980s, the success in breeding the double-zero varieties gave the cultivation of this crop fresh impetus, which was reflected in the cultivation figures. Yields could also be improved and currently average 33 dt/ha. In the case of sugar beet, too, major headway has been made in production engineering. Table 2 provides a comparative overview of the yield development of various crops in the old and new Laender.

Table 2:
Development of yields of key crops in the old and new Laender (and in the former GDR)

	∅ 1959-61	∅ 1969-71	∅ 1979-81	∅ 1987-89	% increase from 1959-61 to 1987-89	∅ 1991-93	∅ 1997-98
	Dt/ha					dt/ha	
New Laender (and in the former GDR)							
All cereals	24.5	30.3	36.5	43.5	+ 77.2	49.4	60.2
Oleaginous fruits	13.9	17.2	20.9	26.1	+ 88.5	26.6	32.6
Sugar beet	233.4	272.2	284.0	290.2	+ 24.3	409.5	460.6
Old Laender							
All cereals	28.7	36.7	44.2	54.7	+ 90.9	60.5	66.3
Oleaginous fruits	21.6	22.4	25.4	31.6	+ 46.2	30.2	32.4
Sugar beet	353.4	445.5	499.5	513.2	+ 45.2	548.9	544.8

Source: Federal Statistical Office; GDR statistical yearbooks; own calculations.

Vegetable production In 1998, a total of 105,435 ha of vegetables, strawberries, and other garden crops were planted in Germany (i.e. 0.9% of the arable land), with 22.4% of this area being cultivated in North Rhine-Westphalia. From 1995 to 1997 the average yields of field vegetables totalled 294 dt/ha. In the EU, Germany ranks fifth in vegetable production, after Spain, France, UK, and Greece.

Viticulture The area under vines in Germany accounts for over 100,000 ha, mostly in Rhineland-Palatinate (66,000 ha) and Baden-Württemberg (25,000 ha). Around 80% of the area under vines is cultivated with white wine varieties, the remaining 20% with red wine varieties and mixed varieties. The yield of white must accounts for over 80 hl/ha, that of red must for around 90 hl/ha, with yields showing marked differences, however, depending on the variety. They are, therefore, not only subject to the usual annual fluctuations, but also depend on the scope of cultivation of the respective varieties. The key white must varieties are Müller-Thurgau and Riesling, Burgundy and Portugieser prevail in the case of red must.

Fruit production   The share of fruit plantations, with 72,013 ha, amounts to only 0.4% of the utilised agricultural area in Germany, but wide regional variation exists. The largest share of UAA in fruit plantations is in Baden-Württemberg with 32%, followed by Lower Saxony with a 15% share. The apple is the key fruit grown in Germany, followed by cherries, plums and prunes ranking third.

Hops   Germany has the largest hop capacity worldwide with a hop-growing area of 17,853 ha (for comparison the USA has 14,841 ha). The hop-growing area worldwide totals 60,112 ha (1998). In the past two years the global hop-growing area has declined by over 10% due to surpluses on the global hop market and the ensuing decline in prices. In Germany the area under cultivation has declined since 1998 by 3,528 ha or by about 16.5%. With a 85% share of the German cultivation, Bavaria is the principal cultivation area for hops, far outdistancing Baden-Württemberg with 8%. Key regions in the new Laender are Saxony-Anhalt (3% of the German area), Saxony (2%), and Thuringia (2%). The expected total harvest for 1999 is estimated at 29,000 t or around 37% of the predicted global hop harvest of 79,250 t.

Tobacco   Tobacco cultivation may not be associated with Germany. There are, however, areas where tobacco contributes a major share to the gross value of farm production. In the 1980s, tobacco was cultivated in the old Laender on around 3,175 ha. After the unification, it temporarily rose to 3,740 ha only to fall again. In 1998 the area under cultivation in Germany totalled 3,847 ha of the total EU area of around 146,000 ha, most of which is concentrated in Spain and Greece. The harvest averaged around 24.7 dt/ha from 1996 to 1998. The area under cultivation averages at 3.7 ha per farm.

Livestock Production

Livestock farming and the production of livestock products plays an important role in German agriculture; livestock products account for around two-thirds of the value of agricultural production and sales, and crop products for only one-third. With respect to livestock, some changes can be noted since 1990 as there has been a marked reduction in the number of animals kept, in the new Laender especially cattle and pigs.

From 1990 to 1997, the number of bovine animals in Germany decreased by an annual average of 3.5% to 15.2 mio. animals. The provisional figures for 1999 show an even further decline to below 15 mio. cattle. In the case of pigs, the population declined annually by 3.1% on average until 1997, and has now stabilized at around 26 mio. Poultry flocks, by contrast, decreased until the mid-1990s, but have now returned to early 1990s levels. Of EU Member States, Germany has the second highest number of cattle (after France), yet the highest number of dairy cows and pigs. Around 18% of the cattle, 23% of the dairy cows, and around 20% of the pigs in the EU are kept in Germany. Sheep are of more importance in the southern regions of the EU, thus less than 2% of the sheep are held in Germany. With respect to poultry, France, followed by UK, Italy, and Spain have the highest numbers; Germany ranks fifth.

Table 3:
Development of livestock in Germany since 1990

Year	Cattle	Pigs	Sheep	Goats	Horses	Poultry
	In thousands					
1990	19,488	30,819	3,239	90	491	113,879
1992	16,207	26,514	2,386	88	531	104,014
1994	15,962	24,698	2,340	95	599	109,878
1996	15,760	24,698	2,324	93	652	112,508
1997	15,227	24,795	2,302	123	b	b
1999[a]	14,819	25,784	2,621	b	b	b

[a] *provisional,* [b] *no surveys*
Source: *Ministry of Agriculture in Germany (ed.): Statistical Yearbook on Food, Agriculture and Forestry, several editions.*

Cattle herds have changed little in the old Laender. In 1960, there were 12.9 mio. bovine animals; in the 1980s, over 15.0 mio., only to fall again after the introduction of the scheme cutting dairy production in 1984. In 1997 12.4 mio. animals were counted. In contrast to cattle, the pig population and sheep flocks rose considerably in the old Laender. The number of pigs has almost doubled since 1950 from 10.1 mio. to 21.5 mio. The number of sheep rose from 1.2 mio. (1954) to 1.6 mio. Goats are important mainly for amateur stockmen as are horses. In the case of poultry, flocks reached a maximum of over 100 mio. animals at the beginning of the 1970s in the old Laender, and have now decreased to slightly over 80 mio. animals. This is due to marked performance increases in both fattening and egg production.

In the former GDR, the cattle population rose from 3.6 mio. to 5.7 mio. from 1950 to 1980. There were then no major changes in the stock size until 1989. The number of pigs more than doubled from 1950 to 1989, from 5.7 mio. to 12.0 mio.. The highest stock numbers were noted in the mid-1980s with over 13 mio. animals. With 2.6 mio. animals, sheep played a far more important role in the former GDR than in the old Laender. In the 1980s, poultry flocks in the former GDR numbered around 50 mio. animals.

What is remarkable for Germany as a whole is the regional distribution of animals. Compared with the livestock in the former GDR, the number of animals markedly declined in the new Laender and is relatively low compared with the national animal population. Thus only 18.3% of the 15.2 mio. cattle in Germany are in the new Laender, and only 13.4% of the 24.8 mio. pigs. In the case of poultry the share is at 26.7%. This means that in considering the share of the UAA in Germany in the new Laender (32.3%), livestock is more heavily concentrated in western Germany than in the new Laender. This may be partly due to the difficult capitalization after the collapse of the GDR, preventing investments in capital-intensive livestock farming sectors.

The performance indicators in livestock production show highly significant growth rates in the past few decades. The development of average milk yields per cow in the old Laender illustrates this best: increasing from 2,498 kg in 1950 to 3,800 kg in 1970, and finally to 4,884 kg in 1990. Thus, the milk yield almost doubled in 40 years. In the same period, the milk yield increased two and a half times in the GDR, albeit from a far lower point of 1,655 kg per cow (1950). A milk yield of 4,180 kg per cow was reached before unification. The milk yield increased considerably in the unified Germany, totalling 5,625 kg per cow by 1997. Fat content lies at 4.25% and protein content at around 3.4%.

In the case of poultry, egg production per hen is a key performance parameter. In 1970 an egg production of 216 eggs per hen per year was reached in the old *Laender*, increasing to 242 ten years later. In the unified

Germany, egg production totalled 260 eggs per hen in 1992, rising to 269 until 1997.

Structure of Agricultural Holdings

Developments in Size and Numbers of Farms

In 1998, there were 516,303 farms with more than 1 ha of farmland, and 27,586 very small farms or intensive livestock farms with less than of 1 ha of UAA, for a total of 543,889. The majority of farms (44.5%) are 1 ha to 10 ha in size, 28.1% are 10 to 30 ha, 12.5% 30 to 50 ha, 10.4% 50 to 100 ha, and only 4.5% of farms are larger than 100 ha.

In the EU, Germany ranks fifth in number of farms, but third in UAA. Thus, the average farm size in Germany is bigger than the average farm size in the EU. In 1997, the average farm in the EU had 18.4 ha, ranging from 69.3 ha in the UK to 4.3 ha in Greece, and Germany recorded an average farm size of 32.1 ha per farm in that year.

The evolution of farm size in Germany is mainly a result of history. In districts where gavelkind (equal division of farm property among all descendants) tenure was practised (southern and southwestern Germany), small-scale farms prevail even today, whereas in regions where farm property was transferred undivided, fewer but larger farms can be found. In East Germany, by contrast, the nationalization of farms during the time of the socialist government has left its mark on structural development there with very large farms even ten years after unification. In a differentiated structural analysis, the districts of gavelkind tenure would therefore have to be considered separately from the regions of undivided inheritance, and the new Laender again would also have to be considered seperately. Despite these historically grown structural differences, however, the overall number of farms has decreased in the old Laender and increased in the new Laender. Therefore, in the following, only a general distinction between the old and new Laender is made.

In the old Laender the number of farms with more than 1 ha of UAA has dropped since 1949, from over 1.6 mio. to 484,290 in 1998. These changes were uneven and fluctuated from year to year with extreme levels of up to 6.4% at the beginning of the 1970s to only 1.5% in 1983 (see figure 1). An examination over time makes clear that the structural change was highest in the decade from 1965 to 1975 with an average annual decrease of 3.2%. The decade from 1975 to 1985 was a period with relatively low decreases in the number of farms, averaging 2.2% per year. In the subsequent decade, structural change intensified again with an average annual decrease in farms of 3.1%. Since the agricultural reform took effect in 1993, the number of farms decreased annually by 3.0% on average.

Figure 1:
Annual decrease rate of agricultural holdings in the old Laender since 1965

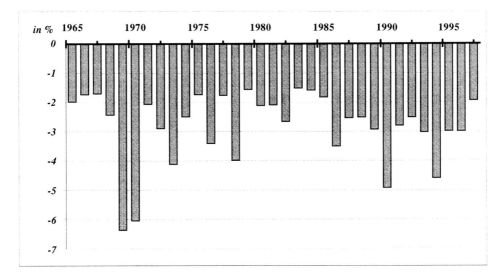

Source: Federal Statistical Office; own calculations.

The growth threshold, i.e. the size class below which the number of farms decreases and above which the number of farms increases, has steadily risen in the past. The farms sized 15 to 20 ha increased until 1969 and decreased as of 1970, meaning the growth threshold for this farm size was reached in 1969. In 1971, the growth threshold was reached for farms sized 20 to 25 ha. At the beginning of the 1980s, the threshold for 30 ha farms was reached and for 40 ha farms, at the beginning of the 1990s. The 60 ha threshold was exceeded in 1997.

The shift in the growth threshold means that despite the decline in the total number of farms, the number of large farms increases to the detriment of small farms. Figure 2 illustrates the relative increase in the number of farms sized 50 ha and more and the reduction in the number of smaller farms of under 50 ha in the old Laender since 1991.

Figure 2:
Development of the number of agricultural holdings according to size classes in the old Laender (1991 = 100)

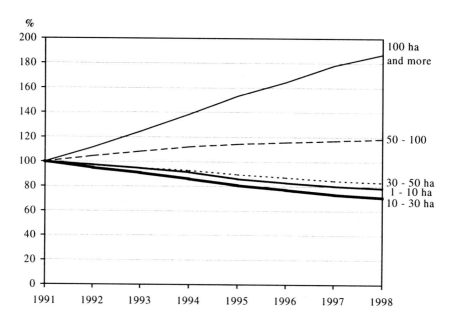

On the other hand, in the new Laender a contrary development can be observed. Collectivization and industrialization of agriculture by the socialist GDR government led to a compulsory reduction in the number of farms and the creation of large farm units. In 1989 there were 8,668 holdings left, 59% of which were run as cooperatives and state farms. They farmed around 90% of the area and owned around 90% of the livestock. The average size of these farms was 1,084 ha. The farms specializing in crop production had 4,547 ha UAA on average under cultivation.

The major restructuring process of agricultural holdings due to unification took place in the first half of the 1990s. Cooperatives of crop and livestock production were merged with an ensuing reduction in some cases in their size as well as in legal form. New farms were also established, mostly as individual farm enterprises. Since 1991 the total number of farms over 1 ha UAA almost doubled from 18,566 to 32,013 (1998). In the last few years, the increase in the number of farms has slowed down.

Figure 3:
Development of number of agricultural holdings by size class in the new Laender (1991 = 100)

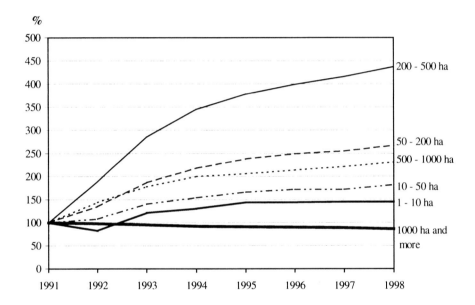

The largest growth rate was recorded in 1993 when the number of farms increased by 36.4% compared with 1992. The growth rate has levelled off considerably since. In 1998 the number of farms increased 2.4% from the 1997 level resulting from the decrease in large farms of over 1,000 ha. Figure 3 illustrates the relative development in the number of agricultural holdings in the new Laender since 1991.

Changes in the number of farms are related to the change in the size of farms. In the old Laender the average farm size grew as the number of farms decreased, this however on the basis of a comparatively low level all in all. Thus, in 1991 farms in the old Laender managed only 19.1 ha UAA, rising to 24.1 ha UAA per farm in 1998. By contrast, in the new Laender, the average farm size has decreased from 284.5 ha UAA to 175.0 ha UAA since 1991, due to the emergence of new farms and a changed farm structure as the large cooperatives have dissolved. Yet these farms are still seven times as large as the farms in the old Laender. Therefore, harmonization of structural conditions between East and West can not be expected in the foreseeable future.

Legal Form and Socioeconomic Conditions of Farms

Differences in the legal forms of agricultural holdings have increased especially since German unification. Whereas in the old Laender family farms were traditionally run as individual farm enterprises, the issue of legal form was relatively difficult in the new Laender, not least due to problems in settling property rights. The following legal forms are distinguished:

- individual farm enterprises, where an actual person is the sole owner (mainly family farms)
- partnerships where several persons are involved (e.g. nontrading partnership, partnership, or limited partnership)
- farms owned by legal entities (e.g. registered cooperative, limited liability company, public limited company, foundation, or central, regional, or local authorities).

Throughout Germany the prevailing legal form among agricultural holdings is the individual farm enterprise with a 97% share, followed by partnerships and legal persons with shares of 2% and 1%, respectively. The latter two legal forms play a great role in the new Laender as they are essentially based on the cooperatives of the former GDR. This also has an effect on the differences in size of the respective legal forms in the east and west. Whereas 44% of farms owned by legal entities in the old Laender are on a smaller scale, under 5 ha UAA, such farms constitute merely 6.5% of holdings in the new Laender. In contrast, over 80% of the legal entities in the new Laender farm a UAA of over 100 ha. Similarly, the partnership farms in the new Laender tend to be larger than those in the old Laender. In the case of individual enterprises, differences in farm size are not as extreme, even if they average more than double the size in the new Laender (49 ha UAA) as in the old Laender (only 23 ha) (see table 4).

Table 4:
Agricultural holdings by legal form (1998)

Legal form	old Laender			new Laender			Germany	
	Number of farms	UAA in 1,000 ha	Ø farm-size in ha	Number of farms	UAA in 1,000 ha	Ø farm-size in ha	Number of farms	UAA in 1,000 ha
Individual farm	472,991	11,036	23	25,925	1,278	49	498,916	12,315
Partnership	8,086	502	62	3,064	1,277	417	11,150	1,779
Legal entity	1,925	93	48	3,008	3,046	1,013	4,933	3,139
of which								
Cooperatives	122	9	76	1,218	1,745	1,432	1,340	1,754
LLC	371	15	41	1,560	1,207	773	1,931	1,222
Total	483,002	11,6327	24	31,997	5,601	175	514,999	17,233

Source: Federal Statistical Office; own calculations.

The socioeconomic conditions are recorded statistically only for individual enterprises. Thus, full-time and part-time farms can be distinguished only for that type of enterprise. Since 1995 full-time farms have been defined as farms with over 1.5 man-work units per farm, or farms with 0.75 to under 1.5 man-work units per farm and a share of the farm income of over 50% in the total household income. All other farms are classified as part-time farms.

Part-time farms dominate with a 59.4% share in Germany. Only 40.6% are run as full-time farms with wide regional variations. In Lower Saxony, Schleswig-Holstein, and the city states (i.e. in the north) over half of the farms are full-time farms, whereas the Laender of Brandenburg, Hesse, and Thuringia show the lowest numbers of full-time farms with a share under 30%. The new Laender in particular have a relatively low share of full-time farms which might be surprising, but this can be explained by the comparatively low stocking density allowing low labour input. In contrast the generally higher stocking density in the old Laender contributes to a more labour-intensive type of farming. Table 5 gives an overview of the share of full- and part-time farms.

Table 5:

Share of full- and part-time farms in individual enterprises in Germany (1997)

	Old Laender		New Laender		Germany as a whole	
	Number of farms in 1,000	Share in %	Number of farms in 1,000	Share In %	Number of farms in 1,000	Share in %
Full-time farms	203.6	41.2	7.6	29.2	211.3	40.6
Part-time farms	290.5	58.8	18.4	70.8	308.8	59.4
Total	494.1	100	26.0	100	520.1	100

Source: Federal Statistical Office; own calculations.

The average utilised agricultural area is far larger in the case of full-time farms (44 ha) than for part-time farms (9.7 ha). Full-time farms in the old Laender cultivate an area of 41 ha UAA, part-time farms 9 ha UAA.

These levels compare to around 127 ha UAA per full-time farm and around 15 ha UAA per part-time farm in the new Laender.

In terms of production activity, 68.5% of dairy farms are full-time farms and 31.5% part-time farms. In the case of pig fattening farms, however, the share of part-time farms is 54.5%, higher than that of full-time farms (see table 6).

Table 6:
Production structure of full- and part-time farms in individual enterprises in Germany (1997)

	Dairy farms[a]			Pig fattening farms[b]			UAA in ha per farm
	Number of farms in 1,000	Share in %	Dairy cows per farm	Number of farms in 1,000	Share in %	Fattening pigs per farm	
Full-time farms	120.2	68.5	29.4	62.5	45.5	101.9	44.0
Part-time farms	55.5	31.5	9.6	74.8	54.5	18.3	9.7
Total	175.6	100	23.2	137.3	100	56.3	23.7

[a] without nurse and suckler cows
[b] animals 50 kg or more liveweight, including culled breeding stock
Source: Federal Statistical Office; own calculations.

The dairy herds kept by part-time farmers average with 9.6 cows per farm, well under the 29.4 dairy cows per farm for full-time farms. Similarly in pig production, part-time farms (18.3 fattening pigs per farm) have smaller herds than full-time farms, keeping 101.9 fattening pigs per farm.

Property Relations

The progressive structural change has not only had an impact on the development of farm sizes, but also on property relations. This manifests itself especially in the number of farms with tenancy land decreasing less strongly than the total number of farms. Due to the great differences between the new and old Laender, separate examination is appropriate.

In 1979 in the old Laender, 57% of farms managed tenancy land, this share, however, rose to 62% in 1997. In the same period, the share of leased UAA in the entire UAA rose from 30% to 48%. Relatively large farms accounted for the bulk of tenancy land: in 1997, the share of these farms with tenancy land amounted to 93.8%, and the share of their land that was tenancy land in their UAA (58.6%) was far higher than the mean value of all farms (see figure 4).

In the new Laender 62% of the 32,000 farms acquired additional land through leasing. However, the share of tenancy land in UAA is much higher (91%) than in the old Laender (48%). Most of the tenancy land (96%) is managed by farms with over 100 ha UAA. On average, farms with tenancy land in the new Laender manage 254 ha, whereas the average tenancy land in the old Laender totalled 18 ha per farm. This high share of tenancy land in UAA is attributable to the particular development in the new Laender. On the one hand, land was returned to the original owners after the fall of the Wall. These are mainly very small areas as expropriations in the former GDR prevented a process of consolidation that occurred in the West. On the other hand, the co-operatives, formed after the fall of the Wall, mostly leased land from their members, as they were unable to draw on previously cooperative property.

In the old Laender, the share of purely tenant farms amounts to 10%, in the new Laender to 27%. These farms manage 11% of the UAA in the old Laender, and 35% of the UAA in the new Laender. These are relatively large farms in the new Laender, whereas they hardly differ in size from the partial tenant farms in the old Laender. The average size of purely tenant farms in the new Laender is 203 ha.

Structure of Labour Input

In 1997, the labour force employed in agriculture amounted to 1.316 mio. persons in Germany on either a full- or part-time basis. Since 1991 around 563,000 persons abandoned agriculture, a 30% decrease. This primarily affected the new Laender due to the restructuring after unification, where an above-average share of permanent employees were

Figure 4:
Farms with tenancy land (1997)

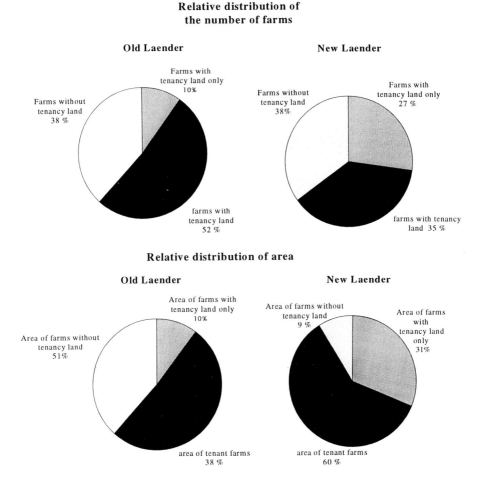

made redundant. Overall, job cuts hit female employees harder than male employees in the new Laender.

To assess the labour input in agriculture, a standardized measuring unit is used assuming a highly efficient full-time employee. This labour input is termed MWU (man-work unit). As a result of higher mechanization, a declining number of farms, and the replacement of labour-intensive with more labour-extensive production sectors (e.g. cereals instead of root crops) the MWUs used has dropped considerably. In 1970, the labour input only amounted to 1,525,500 man-work units in the old Laender, in 1980 to 986,000 man-work units, and only 748,700 man-work units in 1990. Currently only 534,800 man-work units (1997) are being used. A decline in farm labour output was also noted in the new Laender. From 1992 to 1997, labour engaged in agriculture decreased from 173,900 to 115,600 man-work units even though the number of farms increased considerably over the same period. Furthermore, the decline in man-work units in the new Laender has been comparatively higher than in the old Laender as shown by the falling share of man-work units in total man-work units in Germany (see table 7).

Table 7:
Farm labour output in MWU-units, in 1,000

Year	Old Laender	Old Laender in % of Germany	New Laender	New Laender in % of Germany	Germany
1992	672.8	79.5	173.9	20.5	846.7
1993	646.0	81.5	146.3	18.5	792.3
1994	609.0	82.2	132.3	17.8	741.3
1995	571.1	81.8	127.3	18.2	698.4
1996	546.0	81.3	125.7	18.7	671.7
1997	534.8	82.2	115.6	17.8	650.4
1998	512.0	82.4	109.0	17.6	621.0

Source: Federal Statistical Office.

The reasons for the widely varying development lie, inter alia, in the fact that the stocking density in the new Laender is lower than in the old Laender and that more labour-extensive arable crops are being planted (lower share of root crops), so that a lower labour input is generally required in the new Laender. Yet, it must also be considered that the disproportionally strong decline in the labour force in the new Laender was also due to the fact that a relatively large number of employees engaged in agriculture were to some extent assigned to off-farm tasks in the GDR. After unification, farms in the new Laender focused on farm tasks only, thus setting free a large number of employees at the beginning of the 1990s.

Agriculture in Germany – Production and Structure

The employment systems in the old and new Laender show completely different shapes in their basic forms. In the old Laender, the family employment system dominates as the key basic form. Accordingly, farm families accounted for around 85 % of the work force, mainly on a part-time basis only.

In the new Laender, however, the external (non-farm) labour system, employing predominantly non-family employees, has been the key basic form since unification, with a share of nearly 70 %, leaving a share of below 30 % for family members working on the farm.

What is remarkable is the shift in the form of the employment system which can be noted since 1991. Due to the structural change, farms in the old Laender have become larger and increasingly specialized. At the same time, changes in the way farmers and their families see themselves results in the increasing use of non-farm labour to the detriment of family labour. In the new Laender, on the other hand, the share of non-farm labour has slightly decreased as the number of individual enterprises with family labour has increased.

Summary

Around 80% of German land area is used for agriculture and forestry, contributing to less than 1.3% of the gross value added and engaging 2.9% of the total work force. The average farm size varies enormously in Germany, depending primarily on historical conditions. Currently, average farm size ranges from 175 ha in the new Laender to 24 ha per farm in the old Laender. Further changes in the farm structure are expected, especially in the old Laender where the size of farms is still increasing.

Livestock farming is of particular importance to German agriculture. Livestock products account for around two-thirds of the value of German agricultural production and sales proceeds; crop products for only around one-third. In total utilised agricultural area Germany ranks third in the EU after France and Spain, with 17.4 mio. ha.

Chapter 3

German Agricultural Policy – Issues and Challenges

Volker Appel
Federal Ministry of Food, Agriculture and Forestry, Bonn[1]

Objectives and Conditions for Action of German Agrofood Policy

The beginnings of agricultural policy in the Federal Republic of Germany were still determined by the supply bottlenecks during the Second World War and the destitution and food shortages marking the early postwar years. West German agriculture was shaped by a high labor force per unit of area.

In 1955, after years of discussion, the Act on Agriculture was adopted. Section 1 of the Act reads as follows: 'To secure agriculture a share in the progressive development of the German economy and to ensure the optimum food supply to the population, the instruments of general economic and agricultural policies – in particular the commercial, fiscal, credit and price policies – shall enable agriculture to offset its natural and economic disadvantages towards other economic sectors and to increase its productivity. In the process the social conditions of those engaged in agriculture shall be aligned with those of comparable professional groups.'

As a result of a policy of Western integration, i.e. close political, economic, and defense policy cooperation with the states of Western Europe and North America, the Federal Republic of Germany became a founding member of the European Economic Community in 1957. This Community

---

[1] The contribution reflects the views of the author.

also intended to pursue a common agricultural policy with the following objectives as described in Art. 33 of the EC Treaty (former Art. 39, EEC Treaty):

- to increase agricultural productivity by promoting technical progress and by ensuring the rational development of agricultural production and the optimum utilization of the factors of production, particularly labor
- to ensure a fair standard of living for the agricultural community, in particular by increasing the individual earnings of persons engaged in agriculture
- to stabilize markets
- to assure availability of supply
- to ensure that supply reaches consumers at reasonable prices

These two policy statements underscore two priority objectives: assuring the availability of food supplies to the population while ensuring a fair standard of living for the persons engaged in agriculture. The objectives apply to the domestic market. The increase of agricultural productivity was stressed as an intermediate objective or a means to achieve the two objectives. The policy statements laid down as a legal text have not been changed ever since. Yet, in later decades under changing framework conditions, new priorities were added that played a large role in guiding current German agricultural policy.

It soon became clear that the sustained economic growth was accompanied by structural changes in the economy. These led to an outmigration of labor from agriculture to the manufacturing industry and the service sector that continued well into the 1970s. On the supply side, labor productivity increased at a rate of 6% from 1950 to 1998 resulting in declining manpower requirements as the quantitative demand for agricultural goods soon no longer kept pace with this growth in productivity. These changes had a considerable impact on the social security of those engaged in agriculture, in particular on their old age security. Over time the social policy measures accompanying the structural changes in agriculture became an essential part of German agricultural policy.

In the 1970s, the production increases in German and European agriculture, caused mainly by an agricultural support system geared to the internal market, brought growing surpluses of many agricultural products, which could frequently be sold on third country markets only with the help of export subsidies. The ensuing trade policy problems gave rise to the objective of using German agricultural policy to contribute to an improvement of foreign agricultural trade relations. The necessity of coherence of development cooperation and trade and agricultural policies of the EU and its Member States towards developing countries gained in importance, with a view to improving the world food situation.

Due to technological developments and economic framework conditions, certain adverse environmental effects of agriculture surfaced and were increasingly a concern because of changing societal values. Since the 1970s, the contribution of agriculture to ensuring and developing life-sustaining systems has gained stature, and today is an important aim of agricultural policy.

The conditions for action in German agricultural policy have thus undergone changes over the past decades. Germany's reunification raised the issue of a model for agricultural policy in a fundamentally different way. East German agriculture, which was characterized by completely different structures, had to be integrated into the European agricultural market, an endeavor which succeeded due to extensive special provisions, that have since largely expired. Today, the framework for action also encompasses the increasing interdependence of the world economy, the progressive integration of Europe, and the global challenges of sustainable development as set out in Agenda 21. The efforts of consolidating the public budgets of EU Member States as well as the accompanying stringent limitation of public expenditure at the EU level have resulted in severe restrictions on agricultural policy. The current agricultural policy of the German federal government is targeted at a sustainable, multifunctional, and competitive agriculture. At this relatively high level of abstraction there is also a consensus at EU level regarding the further development of the so-called European model of agriculture.

From a national perspective, competitiveness is initially geared to competitiveness of German agriculture in the common market (EU), but also to the increasingly global competition with the gradual liberalization of agricultural markets.

Apart from its competitiveness in the production of food and raw materials, the question how agriculture can meet social requirements will also determine its prospects. In the process, the conservation and shaping of attractive cultural landscapes, the protection of nature and environment as well as a livestock management meeting species-specific needs are playing an increasingly important role. These requirements reflect the changed quality awareness of consumers, who not only make high demands on product quality and food safety, but are also interested in knowing the conditions of production.

The term multifunctionality designates a type of agriculture meeting the above expectations and contributing to employment and economic strength in rural areas. These are often services formerly rendered by agriculture as joint products without financial compensation. Agricultural policy is charged with securing the framework conditions, which will ensure the provision of these services also in the future. This also reflects a more pronounced consumer orientation of agrofood policy.

The demands of society have to some extent been embodied in legal framework conditions (standards for consumer and environmental protection and animal welfare), which in turn exert an influence on the competitive position of German agriculture. The federal governments agricultural policy therefore strives for harmonizing the EU framework conditions relating to competition as far as possible at a high level, where this has not already happened.

The model of the federal government's agricultural policy are farms with sustainable management:

- Economically speaking they strive for a market oriented, cost effective management using modern production processes to as to be competitive on the markets for food and raw materials or to render the services requested by society.
- Ecologically speaking they are marked by environmentally sound management. This includes, in particular, reduced emissions from pollutants, improved energy efficiency, high soil fertility, closed material cycles when possible, species protection, and livestock husbandry meeting species-specific needs.
- In social terms they are to ensure incomes that secure the livelihoods of farmers and farm laborers on their own responsibility.

Areas of Agrofood Policy Activity

Agricultural Market Policy

In 1950 national market organizations were introduced in the Federal Republic of Germany for key agricultural products by creating import and storage agencies and adopting import quotas and tariffs. This served to protect the rebuilding production.

After the foundation of the European Economic Community, Germany embarked on transferring the responsibility of regulating agricultural markets to the Community. A system of European market organizations was developed. These comprised a spectrum of measures that varied according to product and ranged from border protection (variable import duties for a long time), to internal protection (price and sales guarantees by market intervention agencies buying at intervention prices), to the granting of direct production aids (initially only for market organizations without border and internal protection). Since 1962 common market organizations have gradually been established and the common agricultural market was completed in 1967. The Community-level of guaranteed prices for cereals, which operated as pivotal prices, was particularly controversial.

These instruments of agricultural market policy were designed to provide indirect support to the incomes of persons engaged in agriculture by maintaining prices for important agricultural products above the level which would have been set by free trade. With respect to the income objective, this policy had only limited success. Yet, the side effects ensuing from this price policy were crucial. The European Community became a net exporter of important agricultural products. The accruing surpluses led to considerable costs for public storage in the market organizations with intervention schemes. Furthermore, structural surpluses, i.e. not only those that emerged seasonally, could only be sold on the internal market with the help of public price-cutting measures (e.g. aids for utilizing butter and skimmed milk powder) or on third country markets through so-called export restitutions. Coupled with falling customs revenues for the Community, these effects rapidly increased the budgetary expenditure for agricultural market policy.

Agricultural policy responded to this development by introducing further supply-controlling instruments. In 1984, a scheme of milk quotas was introduced on the milk market with the milk producer paying a prohibitively high charge in the case of quota overruns. In economic terms this scheme, introduced in Germany at an individual farm level, operates like a system of production quotas. With the so-called 1988 stabilizer decisions, the EC Agriculture Ministers introduced further instruments to curb production, but soon these also proved insufficient.

In 1986 a new round of negotiations was launched to liberalize world trade within the framework of the world trade agreement (GATT). Whereas previous negotiation rounds largely left agricultural trade aside, they now strive for a gradual liberalization of the sector in the long run. The compromise reached in 1993 (and signed in the Agreement of Marrakech in 1994) with the negotiating partners in GATT, required an intra-Community decision on the fundamental reform of agricultural market policy. Thus in 1992, substantial cuts in support prices were agreed upon, particularly for cereals and beef, which were implemented by 1995/96. To prevent producer incomes from declining, producers received direct compensatory payments linked to area or livestock, which were proportionate to the cuts in support prices.

The new system also had supply-controlling components:
- The area for which compensatory payments (for cereals, oilseeds, and protein plants) are granted is restricted for each Member State. Additional restrictions on cultivation might result from the Blair House Agreement concluded with the United States. This agreement was a necessary step in the search for compromise in the GATT negotiations.
- Each farm (except for small producers) is obliged to set aside a specific share of its so-called base area and not use it for cultivating food or feed crops.
- A specific ceiling was placed on the number of premium rights for young male cattle.

According to the criteria set out in the Agreement on Agriculture signed in Marrakech, these supply-controlling components allowed the classification of the reform-induced compensatory payments into a specific category of measures, the 'blue box', which is exempt from the dismantling process. This is especially significant given that in the Uruguay round the members of the newly established World Trade Organization (WTO) committed themselves to partially dismantling import protection and support measures (internal support, export subsidies) on agricultural markets over a six year period (1995-2001), with special provisions applying only to developing countries.

While the 1992 CAP reform did not comprise all important products, it marks the principal turning point in the Common Agricultural Policy (CAP) as the new direct payments were decoupled from the development of yields. Consequently in the case of the reform products, income policy objectives should no longer be reached via support prices. The transition to a system of direct income transfers increased the administrative burden for all concerned as hundreds of thousands of applications for aid from German farmers alone now had to be processed and spot checks carried out. This placed a particularly heavy strain on the administrations of the Laender.

The reform exerted the expected effects in some respects. The use of domestic cereal grains for feed increased noticeably as a result of their improved competitiveness compared with imported feedingstuffs such as tapioca and soybean meal. (This was, too, a consequence of raised prices e.g. for soybean meal.) In conjunction with set-aside rates of 5 to 15% of farms' base area, this led, at least temporarily, to a definite easing on the cereals market. However, by the end of the economic year 1998/99, cereals intervention stocks had again risen to around 18 million tonnes. In the case of beef the impact of the CAP reform was eclipsed by a slump in domestic demand in 1996, which was in any case declining. This was due to the detection of a possible causal connection between the cattle disease BSE originally occurring in Great Britain and Creutzfeld-Jacob disease in humans. The Council of Agriculture Ministers responded to this market crisis with a whole package of measures for short-term market stabilization to restore consumer confidence and to alleviate the income losses sustained by producers.

In the run-up to the EU enlargement to Central and East European states, in 1997 the Commission presented, together with its opinion on the present applications for accession, its ideas on the further development of Community policies. This text, titled Agenda 2000, also contained reflections on a comprehensive reform of the CAP, in particular the agricultural market policy. The Commission's ideas were based on:

- relatively pessimistic prognoses regarding the development of the international competitiveness of the European agrofood sectors

under status quo conditions (and a consequent drastic rise in EU surpluses and/or unacceptably high set-aside rates)
- the additional need for reforms as a prerequisite for eastward enlargement and as a prior concession for the renegotiation of the WTO Agreement on Agriculture
- stronger consideration of new objectives in the shaping of CAP, particularly environmental concerns

CAP reform soon became one of the most disputed areas of the Agenda 2000, together with the decision on the medium-term shaping of the EU financing system, and reform of Community structural and regional policies. While the compromise reached in the spring of 1999 toned down the Commission proposals on agricultural market policy, key points were adopted in principle. Accordingly support prices for important market organization products were lowered (cereals by 15%, beef by 20%, butter and skimmed milk powder by 15%, the latter as of 2005). These price cuts take place in several annual steps to allow farms to more easily adapt. To mitigate income losses, they are accompanied by a rise in livestock and area premiums. The previously higher oilseeds premium, however, will be cut in three annual steps to the premium level of cereals. The standard obligatory set-aside rate was fixed at 10% until 2006. The milk quota scheme was extended until 2006, combined with reforms of its internal organization and a gradual rise in milk quotas (by 1.5% for Germany).

General provisions for all direct payments in the context of the agricultural support schemes were newly introduced. In the future, all Member States must lay down principles of good professional practice in agricultural technical law or ensure through other measures that the concerns of environmental protection are adequately taken into account (the 'cross-compliance' regulation). Moreover, the Member States are allowed to cut direct payments depending on specific operational criteria (labor input, standard gross margin, premium volume) by up to 20% ('modulation'). The models of an obligatory degression of direct payments, either temporarily or depending on farm size, did not obtain a majority. Further elements of the Agenda decisions relating to agricultural policy will be addressed in later sections.

With the decisions on Agenda 2000, important framework conditions for the development of agriculture up to the year 2006 were adopted. The German side attached great importance to the shaping of financial framework conditions in the negotiations. European agricultural expenditure could be stabilized by confining EU-15 expenditure (for agricultural market organizations and rural development) in real terms to the 1999 level on an annual average from 2000 to 2006. In doing so, the agricultural sector contributes to the strict budgetary discipline. What is more, the Agenda 2000 reform program will not add to the financial burden of Germany, but will actually ease this burden; the German net transfer to the EU budget will drop from a total of 0.55% of the 1999 gross national product to 0.43% in 2006.

The reform continues the process of market and environmental orientation in the agricultural sector, strengthening the competitiveness of domestic agriculture on international markets. The stepwise reform measures and the direct payments will allow farmers and their families to adjust.

The federal government does not deem the compromise on the milk market policy satisfactory since quotas were increased in spite of existing surpluses, and the price cuts are late to take effect. Yet the decisions make it possible to completely eliminate the linking of milk quotas to area. In implementing this option, the federal government has made the rules for transfers of milk quotas more flexible to bolster the economic position of active milk producers.

Agricultural Structure Policy and Policy for Rural Areas

In the late 1960s, a uniform agricultural structure policy developed in the Federal Republic of Germany. In 1969, the Joint Task for the Improvement of Agricultural Structures and Coastal Protection (Gemeinschaftsaufgabe 'Verbesserung der Agrarstruktur und des Küstenschutzes' or GAK) was introduced and has been the cornerstone of German agricultural structure policy ever since. According to the allocation of responsibility set forth in the constitution, the Laender are charged with administering structural policy, including that for agriculture. However, the federal government has coplanning and cofinancing responsibilities within the framework of the above-mentioned Joint Task (of the Federal Government and the *Laender*). The precondition for this is that the tasks are important for the entire population and the co-operation of the Federal Government is necessary for improvement of living conditions (Art. 91a of the Basic Law). The Joint Task was also established to efficiently coordinate agricultural structure policy in Germany and to better harmonize it with the nascent EC agricultural structure policy. The federal government's share of funding in the tasks within the GAK framework usually amounts to 60%. In turn the EU contributes to the federal/Laender expenditure with specific shares, subject to certain conditions.

The statutory objective of the Joint Task is to safeguard efficient agricultural and forestry sectors that are geared to future requirements, while securing their competitiveness and improving coastal protection. In the process, requirements regarding regional and land use planning as well as animal welfare and environmental protection are to be taken into account. Thus, the legislator already expressed that the objectives of the Joint Task should be incorporated in the framework of an integrated development concept for rural areas. The Act on the Joint Task, which has now been amended three times, contains a catalog of measures for an individual holding or for several holdings. A planning committee, consisting of members of the federal (Federal Ministers of Agriculture and Finance) and Laender governments, decides the policy principles and the distribution of federal funds among the Laender. Only the principal measures for the promotion of agriculture and rural areas will be discussed in the following.

The key method for structural adjustment of individual holdings is the promotion of investment by granting aids or low interest loans. In many cases the modernization of farms and growth schemes that have increased productivity would have been inconceivable without policy support. This is particularly true for the structural adjustment process in the new Laender after Germany's reunification. Additionally the promotion of diversification and income combinations (e.g. farm holidays) is important.

To improve marketing structures for agricultural products, support is given to the setup of producer groups (and associations of producer groups) as well as to:
- their general investments
- their investments for specific market partners with whom they have firm commercial relations
- the establishment, development, rationalization, or even closure of marketing establishments in certain sectors, based on the structural plans of the Laender.

The investment support measures also attempted to take into consideration the market effects that could be caused by stimulating capacity-enlarging investments, but this was not always successful.

In the post-war period, the promotion of land consolidation represented the key measure of agricultural structure policy and accounted for the biggest share of GAK support funds well into the 1980s. Its traditional task consisted of raising land and labor productivity by consolidating agricultural holdings (which were due to inheritance traditions, in particular) and creating optimum parcel sizes in terms of labor input. In the meantime objectives have changed fundamentally. Increasingly what matters today is a reconciliation of various economic and ecological claims to land use (e.g. road construction, designation of nature protection areas. These broad tasks mean that land consolidation must be seen as a basic instrument of integral land management.

The promotion of measures relating to water management and agricultural engineering formerly focused on the improvement of agricultural means of production (e.g. construction of rural roads). Today, these measures are primarily designed for inland flood control and the improvement of water quality, e.g. by water retention basins and close-to-nature improvement of water bodies.

A further important measure for individual holdings is the compensatory allowance which is paid as a premium (area-related as of 2000) to ensure the continuation of farming activity in areas disadvantaged due to natural and economic site conditions. This was designed to conserve a minimum population density and safeguard the attractiveness of cultural landscape. Following a substantial expansion of the range of promotional activities, today over half the farmland in Germany is regarded as less-favored, but starting in 2000, compensatory allowance is focussed on especially disadvantaged locations and on grassland.

The fact that GAK has developed into an instrument of rural development also manifests itself in the promotion of village renewal launched in 1984. It helps to maintain and further develop villages as autonomous areas with a decentralized settlement structure. Making villages more desirable places to live, work, relax, and enjoy cultural activities is accomplished, for example, by improving local traffic conditions. Since 1996, the conversion of farm and forestry buildings has also been promoted, frequently creating favorable conditions for the location of business enterprises. As an instrument for rural development, GAK is supported by activities of other policy areas that affect regional development, such as regional economic policy, transportation policy, and labor market policy. In 1999, around 2.8 billion DM of federal/Laender funds were available for GAK, almost 2 billion DM of which went directly to investment promotion.

Agenda 2000 also implied a reorganization of European regional and structural policies. The Regulation on the Promotion of Rural Areas adopted by the Council of Agriculture Ministers combines earlier regulations on agricultural structures and agro-environmental promotion also encompassing (as an integral approach) the promotion of specific off-farm activities, e.g. the promotion of diversification into areas close to farming as well as the encouragement of tourism and craft activities. All in all, the range of measures covering agricultural structures has been substantially expanded and improved. This is a major contribution to the accompanying of changes in agricultural structures, to improving the attractiveness of rural areas with development deficits as well as to creating new alternative employment opportunities. Thus, a Community legal framework for 2000-2006 was established, the prerequisite of establishing rural policy as the 'second pillar' of the CAP.

As part of the integrated policy for rural areas, the Laender drew up development plans covering the period 2000-2006, setting priorities for promotional policy in line with regional requirements. The federal government exerts an influence on the Laender programs, as in all Laender, GAK forms an integral part of programs for rural development. In the future, GAK promotion will focus on measures to improve the competitiveness and ecological soundness of agriculture as well as on measures to strengthen the efficiency of rural structures. In view of the limited funds it will be necessary to regain scope again in promotional policy and to make the necessary funds available also for new, forward-looking measures.

As regards promotional measures to improve the market position of today's agriculture one must take into account the ongoing concentration in food trade. A new aspect is to meet consumer demand for high quality foodstuffs with a regional identity. The promotion of regional production, marketing, and processing chains adopted in the Joint Task in 1999 is appropriate for opening up new prospects for sustainable development in rural areas.

Social Policy in Agriculture

Under the division of responsibilities stipulated in the Basic Law and without a Community program, the federal government is responsible for social security in agriculture. Social policy in agriculture has two important aims:
- to protect against the financial consequences of illness, accident, old age, and long-term care
- to accompany the structural changes in agriculture with a cushion against social hardship.

To achieve these aims, an independent social security scheme for agriculture was set up in Germany to take into account the concerns of self-employed farmers. Those employed in agriculture are, like all employees, included in the general social security systems to protect against social risks.

As far as its core area is concerned the general social security system in Germany dates back to the 19th century. Initially the target group for social policy was only persons in dependent employment. Agricultural accident insurance, which has existed since 1886, has long been conceived as accident insurance for workers. However, since 1939, farmers and farm family workers have been covered by the insurance.

As in the case of all self-employed persons, providing for old age and illness has for a long time been exclusively a matter of making personal provisions. In the case of a farm transfer, the farmer could secure for himself a right of residence and livelihood within the framework of his private pension. The economic and structural developments in agriculture in the post-war period made private provisions for old age only possible to a very limited degree, especially in small- and medium-sized farms due to the rising financial requirements for pensions. As a result the transfer of farms to successors was delayed, which also hampered the spread of innovations. Therefore the oldage benefits, introduced in 1957, aimed at encouraging farm transfers and cushioning them in social terms. These benefits, initially conceived as a supplement to the rights the old farmer keeps when handing over his farm was extended later towards a partial provision. In the 1960s, health insurance for farm families was a controversial subject. The debate unfolded against the backdrop of the alarming state of health of the rural population, the progress of medical technology, and the risk posed by rising health care costs. Since 1972 there has been an insurance liability for self-employed farmers and farm family workers. The further development of the various insurance sectors cannot be documented here in detail. The Act on the Reform of Social Policy in Agriculture ('Agrarsozialreformgesetz'), established, inter alia, independent old age security for farmers' wives. The Act, which took effect in 1995, ended the development of the social security system, at least for the time being.

Already at the beginning of the 1960s, funds from the federal budget were used to finance the special system of social security in agriculture. This use of federal funds was based on the recognition that agriculture, a contracting economic sector, was marked by an unfavorable ratio with too few contributors and too many beneficiaries. The federal assumption of the resulting „old burden", increasing as a result of structural changes, is a primary argument for the use of tax revenue. Thus, the federal Government finances the share of retired farmers' health insurance that is not covered by their contributions. With the 1995 reform, the federal government also committed itself to covering the deficit in farmers' old age security insurance, totaling around 4.4 billion DM in 1999. Altogether the federal government currently provides around 7.5 billion DM per year to finance the social security system in agriculture, considerably easing the burden of social security contributions on farm families. This also demonstrates the great importance of this system in terms of income policy.

Specific social policy measures in agriculture also contain components of structural policy in agriculture. This is illustrated by the farm transfer clause in the agricultural old age security system. It has contributed decisively to the share of elderly farm owners in Germany being well below the EU average. Measures particularly motivated by structural policy were the land transfer pension granted from 1965 to 1983, higher old age benefits in the case of structurally improving land transfers (in particular to farms growing in size), and the similar 'cessation of farming' scheme granted from 1989 to 1996.

Since its beginnings, social policy in agriculture has developed into the most important financial pillar of national agricultural policy. To strengthen the efficiency and competitiveness of German agriculture, in the future it will be necessary to socially cushion the structural changes in agriculture. To continue this task under more restrictive financial framework conditions, more efficient administrative structures are required. Therefore the federal government strives for reform of agricultural insurance institutions. These institutions are the eighty agricultural professional associations, health, nursing care, and retirement funds, which are organized as corporations under public law and are responsible for implementing social policy in agriculture in their districts. In view of the structural changes in agriculture, this large number of institutions is no longer efficient. Furthermore, organizational reform should improve the ability of the federal government to influence the financial and economic management of insurance institutions.

It has been repeatedly questioned whether a separate social security system in agriculture is still required. As a result, the particular features of the separate system were maintained as needed, while approximations were made towards the general system (e.g. the 1995 social reform in agriculture that harmonized benefits and contributions in the farmers' old age security scheme with the statutory pension insurance). The legislation on contributions and benefits as well as the financing system of social security in agriculture are still tailored to the special security needs of farmers.

Up to the recent past, the social security system has continually expanded in Germany. The most recent step, was the improved coverage of the cost risk in the case of long-term care by introducing the independent nursing care insurance in several steps as of 1995. Over the past few years, however, issues of the financing of social security systems and the need for a reform of statutory social security have come to the fore. Several successive acts served e.g. to curb the rising costs for public health, including health insurance in agriculture.

Since federal grants for the social security system in agriculture absorb about two-thirds of the national agricultural budget, change in this field are inevitable as a result of the 'Program for the Future 2000', launched by the federal government in 1999, designed to stabilize the budget and secure jobs, growth, and social stability. Yet, these inevitable savings will not impair the fundamental aims and effects of the use of federal funds for social policy in agriculture.

Agro-environmental Policy

Since agriculture is a sector closely related to the environment, its sustainability and environmental effects are important. Over centuries, agriculture actually created the diversity of cultural landscapes in Germany with their region-specific flora and fauna. Today, pursuing nature conservation and landscape management in Germany would be inconceivable without land management. For example, over 50% of endangered wild animals and plants in Germany inhabit open landscapes and would hardly continue to exist without management activities.

Coupled with economic and institutional framework conditions, agricultural engineering, fertilization and plant protection, plant and livestock breeding, and improved feeding have brought great progress in agriculture. This has resulted in a greater quantity of cheaper food of higher quality and the availability of more renewable resources than ever before. However, since the 1950s such developments and the new economic framework conditions have also led to increasing strain on soils, water, and air. Habitats of wild animals and plants have either been destroyed or impaired. Over the past few years important steps towards a more ecologically sound agriculture have been taken by agricultural policy causing a trend reversal in several fields. Thus, despite increasing yields, sales of commercial fertilizer and the accumulation of nutrients from animal husbandry per ha of farmland have declined substantially since the end of the 1980s. The ecological compatibility of plant protection products is steadily increasing due to the stringent requirements set by the Plant Protection Act.

In spite of these encouraging developments, problems persist. Regionally and locally, substance inputs from fertilizers and plant protection impair groundwater and surface water, particularly by nitrate inputs, posing a serious environmental problem. With manure and mineral fertilizer application, as well as methane emissions from cattle, German agriculture has a substantial share in the emissions of the greenhouse gases nitrous oxide and methane. However, agriculture only accounts for around 6% of total greenhouse emissions in Germany. Agricultural production accounts for 90% of ammonia emissions in Germany, which can cause damage to plants and the acidification and eutrophication of ecosystems. Plowing up grassland and changing the use of land has seriously impaired the habitats of numerous wild flora and fauna in the past.

Agro-environmental policy addresses these problems by, first of all, putting the rules of good professional practice into concrete terms. Good professional practice is designed to draw the line between a method of production society can expect of agriculture without additional remuneration and particular services rendered for nature and the environment, which go beyond this. Key parts of good professional practice are regulated in agricultural technical legislation and in other legal fields relevant to agriculture. On the one hand, they have been shaped as legal provisions over the past few years. The 1998 Federal Soil Conservation Act contains the principles of good professional practice with respect to soil use. The Fertilizer Ordinance, adopted in 1996, specified the principles of good professional practice in fertilization. The amendment to the Plant Protection Act, adopted in March 1998, represented a substantial expansion of good professional practice. Additionally comprehensive, nonbinding principles and recommendations for good professional practice in the fields of soil use and plant protection based on these legal provisions were recently issued. These rules of good professional practice constitute effective environmental minimum standards. Supplemented by bans on the use of individual areas and specific nature conservation measures for areas used for agriculture and forestry, a differentiated land use emerges, as demanded by environmentalists and nature conservationists, which is regarded as economically efficient under the aspect of sustainability.

Beyond the further development of regulatory rules, efforts are required to improve and spread production-engineering measures, in particular to reduce emissions. Agricultural research related to the environment makes key contributions to this, as does the departmental research in the purview of the Federal Ministry of Food, Agriculture and Forestry. Today, research geared to the protection of ecosystems and resources as well as to an environmentally acceptable agricultural production meeting high quality standards is a priority of departmental research. Particular requirements of environmental and nature conservation policy, beyond good professional practice, must be realized through financial incentive measures, which have proved their worth as near-market measures in terms of efficiency as well as acceptance. Following a voluntary extensification scheme adopted at the EU level in 1988, the promotion of ecologically compatible production processes protecting natural habitats became part of the flanking measures in the 1992 CAP reform. Cofinanced with EU and (partly) federal funds, the Laender supported – as part of voluntary agro-environmental programs – extensive production processes as well as measures of biotope and contract nature conservation to, among other things, conserve the cultural landscape. With wide regional variations, these programs covered around 29% of German agricultural area in 1998. The federal government is involved through the promotional principles for 'market and site-adapted land management' in GAK.

Under Agenda 2000, the agro-environmental programs continue as part of the Regulation on the Promotion of Rural Areas. It may be noted that these are the only measures of the Regulation whose application is binding for the Member States. The prospect of rewarding farming methods that are particularly compatible with the requirements of environmental protection were further improved by generally raising the maximum rates eligible for cofinancing. Furthermore, in the future, farmers in certain areas can be compensated if land management is restricted due to environmental and nature conservation requirements based on Community law. The so-called Flora-Fauna-Habitat Directive is a well-known example of such law.

Where voluntary environmental measures do not lead to a satisfactory result for agro-environmental policy, stricter – but compensatory – legal requirements would be appropriate. The claim for such a compensation is laid down in the Federal Nature Conservation Act.

The importance of organic farming has steadily increased in Germany. Yet the share of organic products in the total turnover of food was a mere 2% in 1998. Under the conditions prevailing in central Europe, organic farming (despite the lower yields) meets the sustainability principle in a particular way because of – inter alia – its favorable consequences for the protection of natural resources. Therefore, the federal government will support the further development of markets for organic food through stepped-up research, consumer education, intensification of public relations work, and primarily through improved promotion of the production, sale, processing, and marketing of organic products. In the future, the further expansion of production will be economically viable only if the demand for organic products will increase correspondingly. To this effect, obvious shortcomings in the sale and marketing of organic products must be overcome.

Initially, the production of renewable resources was promoted due to the related easing of the strain on surplus markets. Nowadays, aspects of sustainability play a great role however. To indirectly support demand, the federal government promotes research, development, and demonstration projects. The new market introduction scheme to promote renewable energy sources will give greater impetus to the use of biomass for energy generation. The framework for the use of renewable resources has been improved through legislative measures (e.g. ecotax, Packaging Ordinance, Ordinance on Biowaste and the Electricity Supply Act). In 1999 renewable resources were planted on about 740,000 ha in Germany.

All environmental and resource policy measures should be accompanied by market and other promotional policies supporting the environmental and sustainability aspects. In this respect the 1999 decisions regarding the agricultural chapter of Agenda 2000 created better framework conditions for an ecologically sound agriculture. What will matter in the medium and long term is to use the possibilities of grassland promotion and to contemplate a corresponding conversion of the premium system. The aim is the preservation of open countryside, particularly in less-favoured grassland regions.

Consumer-Oriented Agricultural and Food Policies

Besides the traditional fields of action addressed above, other issues of agricultural and food policy are under discussion in Germany. The discussion focuses on conflicting goals regarding food quality in a broader sense and, in particular, the application of modern technologies in food production (e.g. genetic engineering). With a view to a consumer-oriented agricultural and food policy, the federal government aims to strengthen precautionary consumer protection and better inform consumers. Thus, for example, since 1 July 1999 there has been a ban in Germany on using a number of antibiotic growth promoters as feed additives. This ban was enacted to reserve human medicines so that antibiotics used in human medicine or which trigger cross-resistances with human antibiotics are not allowed as additives. The second example is the existing ban on the importation of beef from hormone-treated cattle. This action, too, gives precedence to consumer protection over economic interests.

The improvement of animal welfare is a central concern of the federal government. The EU Directive on the Protection of Laying Hens adopted in 1999 by the Agriculture Council constituted a considerable step forward in this regard. In the long run, this Directive prohibits the keeping of laying hens in conventional cages and gradually legislates the improvement of conditions for these animals. As German agriculture is exposed to European competition, improvements in husbandry provisions can be made only at the European level. With respect to animal transport, apart from general limits on the duration of transport, strict requirements apply to the loading, feeding, and watering of animals with broad powers assigned to inspection authorities. The federal government supports legislative initiatives in the German Parliament to adopt animal welfare as a constitutional objective into the basic law.

Prospects

The results of the current WTO negotiations as well as the eastward enlargement of the EU will be decisive for future international framework conditions for agriculture and the food industry.

With regard to a further liberalization of world agricultural markets, the federal government places special emphasis on the necessity of fair conditions of competition. During the WTO negotiations, securing a suitable share of world agricultural markets for the European agrofood industry will be sought. On the other hand, the safeguarding of a multifunctional agriculture in Germany and Europe, which produces according to high standards of consumer, environmental, and animal protection, is at stake. Having achieved a stronger market and environmental orientation of the CAP with the Agenda 2000 reform, the EU is on solid footing for the WTO negotiations. Particularly the decisions on arable crops, beef, and the promotion of rural development provide a good starting point.

The integration of the agricultural economies of the Central and Eastern European candidate states into the common internal market is one of the key challenges in realizing the enlargement of the EU. Agriculture plays an important role in these states, if only because of the above average share of people engaged in agriculture. Proceeding carefully and cautiously during integration will be crucial to avoid abrupt changes. Agenda 2000 also well equips the Community for eastward enlargement. Of particular note are the pre-accession aids designed to assist candidate countries with their structural policies and to facilitate their adoption of the *acquis communautaire*.

Chapter 4
Transformation of Agriculture in East Germany

Bernhard Forstner and Folkhard Isermeyer
Institute of Farm Economics and Rural Studies
Federal Agricultural Research Centre (FAL), Braunschweig

Introduction

After the fall of the iron curtain in 1989, the German Democratic Republic (GDR) acceded to the Federal Republic of Germany (FRG) in 1990. All political, economic, and social systems of the FRG were also put in force in East Germany. This caused a fundamental transformation of nearly all parts of East German society.

This chapter focuses on the framework, major features, and results of the transformation process in agriculture of East Germany (New Laender) since the reunification in 1990.[1] The first section describes the historical background which led to the emergence of huge state and cooperative farms in the wake of appropriation and collectivisation of farmland and assets. The second section covers the economic situation and the most important political decisions after 1990 which dominated or influenced further development of the agricultural sector in East Germany. The following section contains an overview of some structural developments. The current economic situation of agricultural holdings in the New Laender is

---

[1] The term 'New Laender' comprises the area of the former GDR after 3 October 1990. Colloquially, this part of Germany is called 'East Germany'. In this chapter, we use both expressions interchangably. Analogously, 'West Germany' or 'Old Laender' refers to the territory of the FRG prior to 3 October 1990.

presented in the fourth section. In the last section we assess the results of the transformation process and provide an outlook on further developments.

Historical Background

Before World War II, the farm structure in Germany was already rather heterogenous. As a result of various factors, especially inheritance practices, small farms dominated in the central, western and southern parts of Germany, whereas larger farms prevailed in the north. Northeastern Germany, especially the trans-Elbian territory, was characterised by very large farms.

At the end of the war in 1945, the Soviet Union took control of East Germany. The Soviet sector comprised about one-third of the German territory and one-fourth of the German population. After 1945 the political and economic system in the Soviet sector (after 1948, the GDR) developed in line with the socialist principles that have been enforced by the Soviet Union and socialist/communist parties all over Central and Eastern Europe.

The economic transformation was started by land reform and the expropriation of industrial property. The land reform, which was carried out between 1945 and 1949, affected about half of the agricultural area in the Soviet sector. All farms with more than 100 ha (7,160 farms, 2.5 million ha), the farms of war criminals and members of the National Socialist Party, as well as 4,537 other farms (132,000 ha) were expropriated without compensation.[2] This land was pooled with state land, and of the total pool (3.2 million ha), 2.2 million ha were then distributed among 559,000 newly established or small farmers.[3] The remaining 1 million ha remained state property (Weber, 1991). This land was used to create state farms which were to be models of modern socialist agriculture. As the newly established private farms were very small (on average 8 ha) and lacked other essential preconditions for efficient farming, 20% of these farms were

---

[2] Some additional expropriations occurred after 1949. Until the end of 1953, a further 24,211 farms, about 700,000 ha farmland, were taken by the state without compensation. For further details see Bell, 1992.

[3] The total of GDR farmland amounted to about 6.2 million ha.

given up by mid-1953 (Weber, 1991). Many of these farmers formed the first agricultural producer cooperatives.

In 1952, the collectivisation of family farms was officially proclaimed. It was implemented in several steps. In the first stage, mainly small farmers formed a producer cooperative on a voluntary basis. In subsequent years, the government exerted more and more economic and political pressure to accelerate this process. Finally, from autumn 1959 to spring 1960, medium-sized farms especially were forced to join the local cooperatives. As an evasion, many farmers abandoned their property and fled to the West. Hence, by the middle of 1960, almost all private farms in the GDR had been transformed into large collective farms. On average, the private farms of one village formed one collective farm (LPG).

In the 1970s, a second period of collectivation was started. The intent was to systematically move to a purely industrial style of agricultural production. This was to be achieved by further specialisation and concentration. From the holdings of two, three, or four collective farms, one big cooperative was established which was responsible for the total plant production including the production of feed on about 4,000 ha. The feed had to be transferred to the other formerly collective farms which continued to exist as pure animal production units. Many specialised service units (e.g. agro-chemical service centers or ACZ) were established in order to connect the large specialised production units. In so-called cooperation councils, the exchange of fodder and other products and services between the plant and animal holdings were coordinated. This led to severe organisational problems. In short, the strict organisational separation of animal and plant production units in the framework of the socialist system led to a high degree of inefficiency.

The political and economic system of the GDR was strictly hierarchical. The means of production as well as output were distributed according to central plans. These were the basis for regional and individual farm plans. Therefore, LPGs were not free in their decisions regarding production and investment. Following the goal of regional autarky, the LPGs often had to implement, against their will, the investment plans issued by the central planning institutions. These investments were frequently connected with high financial obligations to the National Bank for Agriculture and Food Economy. On the other hand, the performance of the large-scale farms was often impaired by insufficient supply and poor quality of input goods (machinery, feed, etc.).

The LPGs were not only large-scale agricultural enterprises. In many rural areas they were practically the only employer, i.e. they offered work in the villages and they dominated social and cultural life. Compared to many family farmers in West Germany, the members of an LPG enjoyed a number of privileges (fixed holidays, workfree weekends, comprehensive social security, etc.). On the other hand, there were few incentives to increase productivity:

- while real property formally remained with the owners, the use of the land was severely restricted and the land had to be let to the LPG without rent
- in the course of time, the collective farmers became the minority in the large-scale cooperatives, thus decision-makers were alienated from the land ownership
- since wages were not differentiated according to performance, motivation was generally fairly low and people concentrated more on individual, small plot farming

The small-scale individual household production of agricultural goods remained or became an important feature of East German agriculture. It is estimated that in 1988 about 15% of the meat, 33% of the eggs and of fruit, as well as 14% of the vegetables had been produced by individual households (BMELF, 1991).

Over time, the price structure for agricultural inputs and products became severely distorted, partly due to the planned levels of self-sufficiency. At the end of the 1980s the ratio between livestock prices and prices for field crops was much in favour of the former. Consumer prices were extremely subsidised so that it was ultimately advantageous to fatten pigs by feeding bread instead of grain. Hence, severe economic distortions and misallocations occurred and led to substantial inefficiencies.

As agricultural production was mainly influenced by state orders without regard for natural and economic conditions of specific locations, in some locations severe environmental problems occurred. In particular the establishment of huge, specialised livestock farms caused major environmental damage because there was no adequate legal framework for pollution control. It is important to note that, despite all the problems mentioned above, East German commentators regarded agriculture as one of the most stable sectors of the GDR economy (Reichelt, 1992).

Table 1 shows the farm structure in the GDR at the end of the 1980s. Altogether 4,530 cooperatives and 580 state farms managed about 90% of the total farmland and livestock. Both, cooperatives and state farms, were specialised in either crop or animal production. One thousand three hundred sixteen holdings engaged in crop production on average farmed 4,000 ha farmland; 95% of these holdings cultivated more than 2,000 ha, 61% more than 4,000 ha, and 20% more than 6,000 ha (Eberhard, 1991). All told, about 825,000 persons or 10% of the total working population, were engaged in farming or related activities.

Table 1:
Key data on agriculture in the GDR, 1989

	Holdings	Employed persons			Agricultural area			Livestock		
		in 1,000	in %	per holding	in 1,000 ha	in %	per holding	1,000 units	in %	per holding
Agricultural cooperatives (LPGs)	4,530	694.9	84.2	153.4	5,075	82.2		4,343	74.5	
crop production	1,164	306.9	37.2	263.7	4,987	80.8	4,284.4	69	1.2	59.3
animal production	2,851	343.6	41.6	120.5	73	1.2	25.6	4,273	73.3	1,498.8
horticultural production	199	27.6	3.3	138.7	15	0.2	75.4	1	0.0	5.0
Agricultural state farms (VEGs, VEBs)	580	124.8	15.1	215.2	464	7.5		971	16.7	
crop production	152	46.2	5.6	303.9	408	6.6	2,684.2	36	0.6	236.8
animal production	312	49.1	6.0	157.4	39	0.6	125.0	443	7.6	1,419.9
other	116	29.5	3.6	254.3	17	0.3	146.6	492	8.4	4,241.4
Total	5,110	819.7	99.3	160.4	5,539	89.8	1,084.0	5,314	91.1	1,039.9
Agriculture, total *		825.2	100.0		6,171	100.0		5,830	100.0	

* In addition to the holdings mentioned there were 375,000 households with small plots and 3,558 other small holdings.
Source: BMELF, 1991.

At that time in the FRG, on an area about twice that of the GDR, there were 648,800 farms with an average size of 18.8 ha farmland. The total workforce in West German agriculture amounted to 1,617,000 persons or 775,200 worker units.

Policy Measures After 1989

In June 1990, only seven months after the fall of the Iron Curtain, the newly elected government of the GDR and the government of the FRG agreed on a basic treaty which paved the way for the unification of both economies. In order to supply East Germans with sufficient purchasing power and to prevent mass migration from East to West Germany, the exchange rate of GDR mark to FRG Deutschmark was fixed at one to one for small amounts of money and at two to one for the rest. By far, the latter ratio did not reflect reality since the shadow ratio was seen as between five and ten to one. On 3 October 1990, the unification of Germany was completed.

Short-term Aid and Transfers Under the Common Agricultural Policy

The unification of Germany was combined with the accession of East Germany to the European Union (EU). After several transitional adjustment regulations, the 'aquis communautaire' applied to the New Laender. The various market regulations had to be transferred to the New Laender in 1991. Among these were milk and sugar quotas which led to a significant reduction of East German milk and sugar production compared to 1989.

The German government and the EU offered massive help to alleviate the problems arising through actual financial bottlenecks and through the necessity to adjust structurally to the new economic situation. Between 1990 and 1995 17,200 million DM were spent to overcome the financial stress due to price reductions for agricultural products, prevent the breakdown of a large quantity of potentially viable enterprises, and establish a modern agricultural sector in East Germany (BMELF, 1996).

Aside from short-term emergency payments, major funding was given to more mid- and long-term structural measures. Although general regulations for structural adjustment of agricultural production and the marketing sector in the EU already existed, special conditions were agreed to for East Germany since its structural problems and backwardness compared to other EU Member States were distinct. East Germany was classified as an EU region with top promotion priority. This meant that up to 75% of all expenses for structural programmes were paid by the respective EU fund. This privilege also applies to other Member States or regions which lag far behind the average EU economic level. To avoid personal hardships due to the necessary cutbacks in the workforce in the LPGs, the federal government offered supporting social security measures (e.g. early retirement, unemployment benefits).[4]

Agricultural Adjustment Law

The amended Agricultural Adjustment Law (Landwirtschaftsanpassungsgesetz) which came into force in July 1991 was basic to the structural development of agricultural holdings (Schweizer, 1994). The foundations of the law were *(a)* warranty of full property rights and *(b)* equal opportunity for the different legal and organisational forms of holdings. The main fields regulated in this law were the following:[5]

- LPGs had to expire by the end of 1991. The holdings could be liquidated or transformed to another legal form of the German commercial law by division, fusion, or simply a 'name' change. The law had important procedural aspects.
- Termination of membership in the LPGs was made possible, and claims in case of exit were determined.
- LPGs were forced to distribute their capital among the members according to certain statutory requirements in order to repay the originally paid asset contributions and the lack of land rent during the years since collectivisation. This was regarded as necessary since the formerly collectivised farmers were now a minority in many of the LPGs.
- The amount of capital had to be determined on the basis of an orderly balance sheet.
- Members who wanted to reestablish individual farms had to submit their claims promptly for full payment whereas other members could be paid in installments.

---

[4] Further details, see Hagedorn and Mehl, this volume.
[5] According to the first Agricultural Adjustment Law, still passed by the last legislature of the GDR in 1990, LPGs could only be transformed without liquidation into a registered cooperative. This regulation and further regulations to the detriment of formerly collectivised farmers led to an amendment in 1991.

During the transformation of LPGs into another legal form, the assignment and distribution of capital in line with the agricultural adjustment law was quite important for the further development of both the LPG-successor farms and the new or reestablished farms. However, the process and the result of equity distribution has been little analysed. The determination of distributable capital was an especially difficult task and left room for maneuvering or shaping the outcome (Thiele, 1998, p. 64 ff.). Even experts did not know exactly how the relevant capital was to be calculated. Hence, law courts had to settle many controversies.

Investment Aids

At the end of 1989, machinery and buildings were in an advanced state of decay. Many of the still usable assets were not in good condititon. After the unification, the scarcest factor of production was capital. LPGs and their successors needed capital to pay off exiting members and to upgrade their own equipment.

As in most cases, where available funds were meager, the state offered financial support. Two major investment support programmes were designed with the special requirements of new or reestablished farms and of LPGs or their successor farms in mind. As a means to foster animal production, livestock-related investments have been subsidised more intensely than investments in crop farming since 1994. Almost all farms had got assistance from one of the investment support programmes by the end of 1996 (BMELF, 1998, p. 84).

The investment assistance favoured newly established individual enterprises, especially partnerships. Hence, the promotion policy had a major influence on the foundation of farms of this type. Partnerships could get up to three times the level of assistance provided to single enterprises. This policy was also applied for the assignment of milk quotas; partnerships could get two times the amount of an individual farm's quota.

Since 1997 public investment assistance in East and West Germany has been made uniform. As West Germany has a long tradition of subsidising farms that invest, this development does not necessarily mean a radical reduction of investment aid for East German farms.

"Old" Debts

A major obstacle for many LPG-successor farms was debt assumed prior to unification. Since these debts were partly caused by unproductive or inefficient investments commanded by the state, the remaining obligation was a heavy burden on many farms as the debts became commercial loans after the unification. Altogether, about 50% of the LPG-successor farms carried debts from before 1990 which totalled about 3 million DM per LPG on average, but with large variation.

In order to avoid numerous bankruptcies, the government provided special treatment for these debts. First, a certain share of the debt for unproductive investments was assumed by the state. Second, the affected farms were allowed to remove the remaining „old" debts from their balance sheets; they could continue farming and were obliged to repay the „old debts" only if they showed a profit. This regulation, and other incentives (e.g., interest rates below commercial terms; a waiving of compound interest, and the right to deduct redemption payments from taxes) meant that none of the firms went into bankruptcy because of existing debt. Since the repayment terms offer room to maneuver, and since interest accumulates slowly due to the fixed favourable conditions, farms could delay their repayment of „old debt" for a long time without consequences. The incentive to show a „regular" profit is low.

The Privatisation of State Farms and Land

The West German government maintains that the Soviet Union regarded the retention of the land reform as one of the decisive preconditions for the unification of both German states. Therefore, both countries stated in their common declaration for the regulation of property rights problems (a part of the unification treaty) that the expropriations carried out from 1945 to 1949 will not be reversed. Even today, this basic decision and its background are very controversial. The former owners of expropriated land brought the case before the Federal Constitutional Court. In two judgements, the court upheld the basic decision of the government.

However, the government did not intend to remain owner of the expropriated farmland. The state-owned farmland, about 1 million ha agricultural area, should be privatised.[6] The conditions of privatisation were intensively negotiated in a time-consuming political process. During this period the land was leased on a short-term basis at first, and later on a medium and long-term basis. The cornerstones of the privatisation were fixed in a special land acquisition programme according to the Compensation Act of 1994 (Entschädigungs- und Ausgleichsleistungsgesetz, or EALG):

- owners of expropriated land are eligible for either a small compensation or for a subsidised purchase of a small amount of land compared to their original areas
- all farms with a lease-contract for state land are eligible to buy such land
- prices of this land are about half of the common market price in the New Laender

The privatisation of farmland was started in 1996. In the meantime the European Commission required the German government to adjust the conditions of land acquisition. After that, the privatisation was stopped and the legislation process was taken up again.

The Development of East German Agriculture Since 1990

Prices, Production, and Workforce

As a consequence of the unification of the two German economies in July 1990, producer prices of all agricultural products in East Germany drastically declined. Since the prices for agricultural products had been much more subsidised in the GDR than in the EU, prices had to adjust to the lower West German level. However, prices for East German products, especially livestock products, fell far below West German levels. The major reasons were (a) a sudden loss of export markets in Eastern Europe, (b)

---

[6] There was also a large amount of forest which was confiscated and was now to be privatised.

takeovers of East German retail chains by West German companies preferring to stock their own products, *(c)* changed consumer preferences in favour of long-desired Western products, and *(d)* poor quality of many processed food products manufactured in East Germany. Financial bottlenecks led to panic and emergency sales which aggravated the price situation.

The price reduction differed considerably with respect to product groups. Crop products were hit much less than animal products. Before 1990, the price gap between East and West Germany was smaller for crops than for animal products. This reflected the comparable quality and marketability of crop products from East Germany.

Milk and meat producers suffered dramatically from the low producer prices in 1990 and 1991. This was mainly due to poor quality and problems in the processing sector. The average dairy milk price in 1991 in the New Laender was significantly below the price in West Germany. In the following years the price gap narrowed, but has not yet closed up. This development is the result of better milk quality and new modern processing factories which are now capable of paying competitive prices.[7]

Investments in processing industries, which were extensively subsidised by public funds, have improved the competitiveness of East German agriculture considerably. There are however, a few exceptions. The high number of slaughterhouses, sometimes built at unfavorable locations, have become a heavy burden for the meat industry, impeding a revival of animal production in East Germany (Wolffram, et al., 1996).

The development of input and output prices was different. A comparison of 1991 against 1989 shows that output and input prices went down on average. However, while output prices decreased by about 65%, input prices dropped by only 20%. Single inputs and outputs showed different changes. While prices for new buildings increased by about 60%, machinery cost about 40% less than in 1989 (BMELF, 1995).

The drastic decline in product prices put many farms in serious financial straits. They tried to remain solvent by selling marketable assets and reducing cattle and pig numbers. From winter 1989/90 to summer 1992,

---

[7] As a consequence, now a major milk flow to East German dairies arises due to the increased prices while in the first years more than 10 % of East German milk production went to West German dairies (ZMP, 1999).

cattle numbers in East Germany declined by almost 50 %; pig and sheep numbers decreased even more.

As the most dramatic changes took place from 1990 to 1992, Table 2 includes data for each of these years and for 1998. For comparison purposes, the table also displays figures for West Germany.

Table 2:
Selected features of German agriculture during transition

Aspects	Region	Unit	1990	1991	1992	1998
Farms > 1 ha	**New Laender**	**number**	**5,110** [1]	**18,566**	**18,609**	**32,013**
	Old Laender	number	648,772 [1]	597,703	581,934	484,300
Workforce	**New Laender**	**persons**	**825,200** [1]	**361,700**	**202,000**	**144,500**
	Old Laender	persons	1,617,000 [1]	1,516,500	1,460,300	1,116,500
Cattle [2]	**New Laender**	**1,000 head**	**5,540**	**3,264**	**2,831**	**2,712**
	Old Laender	1,000 head	14,542	13,869	13,377	12,229
Dairy cows [2]	**New Laender**	**1,000 head**	**1,906**	**1,103**	**1,036**	**954**
	Old Laender	1,000 head	4,771	4,529	4,329	4,838
Pigs [2]	**New Laender**	**1,000 head**	**11,088**	**4,702**	**4,400**	**3,582**
	Old Laender	1,000 head	22,036	21,362	22,114	22,717
Milk per dairy cow	**New Laender**	**kg per year**	**4,260**	**4,432**	**4,919**	**6,430**
	Old Laender	kg per year	4,857	4,942	5,052	5,560
Grain yield (winter wheat)	**New Laender**	**100 kg/ha**	**55.4**	**60.1**	**46.5**	**60.6**
	Old Laender	100 kg/ha	66.6	72.1	67.7	61.2
Milk price	**New Laender**	**DM/100 kg**	**n.a.**	**49.69**	**53.90**	**57.03**
	Old Laender	DM/100 kg	62.56	60.28	60.63	58.38

[1] 1989. [2] December counting.
Source: ZMP; BMELF; StBA.

Together with livestock there was also a radical cut in the agricultural workforce. In 1989 about 820,000 persons, or about 10% of total workforce in the GDR, were employed in LPGs and state farms. Since the large farms were not only engaged in agricultural production, a major share (about 25%) of their total workforce was engaged in nonagricultural activities. Among them were cultural and social tasks, the maintenance of machinery and buildings, and various crafts. Until 1989, the poor state of machinery and buildings had required increasing labour for repairs. Furthermore, a special system of money allocation had created an incentive to employ a large number of persons. Since labour had not been scarce, there were many redundant persons employed in agriculture until 1989.

After 1989, the agricultural workforce was reduced to 362,000 persons in 1991 and 202,000 persons in 1992 (Table 2). By investing in better machinery and using less labour-intensive production methods, the workforce was further cut to 144,500 persons by 1998. Family labour only accounts for 28% of the total agricultural workforce.

The productivity of East Germany's agriculture has increased remarkably during the past nine years. In some subsectors (e.g. dairy and rapeseed), yields are now even higher than in West Germany. Compared to the average of 1986 to 1991, the average yield from 1997 to 1999 of winter wheat and sugar beet rose by 43% and 34% respectively. This increase in crop yields is mainly due to *(a)* the availability of Western technology (seeds, fertiliser, pesticides, machinery), *(b)* incentives created by market forces, and *(c)* a better adjustment of production patterns to natural conditions. The increase in milk yields is predominantly the result of better feed quality, advanced feed management (e.g. total mix ration), and higher animal comfort.

Farm Structure

The transformation of agricultural holdings has been closely scrutinised in various analyses. The major points of interest were the progress of establishment and reconstruction of farms, the main factors of influence, and the prospective structural development in light of existing theoretical findings (Isermeyer, 1991; König and Isermeyer, 1996; Welschof, 1995; Schmitt, 1993).

The number of agricultural enterprises rose from 5,110 cooperatives and state farms in 1989 to 18,600 in 1992, and to about 32,000 in 1998 (Table 3). The speed of transformation decreased markedly after 1992. The main features of the transformation have also changed over the years. In a first step, many LPGs decided to reunify animal and crop production. In some cases however, a union didn't take place and the large specialised crop production holdings remained largely unchanged. In a second step, LPGs had to change their legal status. As prescribed by the Agricultural Adjustment Law, producers' cooperatives had to transform into a legal form according to German commercial law. State farms had to be privatised. Only a minority of the cooperatives decided to liquidate.

Table 3:

Number of agricultural holdings in the New Laender since 1992 according to legal form*

Legal form	Number of holdings				Change in %		
	1992	1994	1996	1998	1994 : 1992	1996 : 1994	1998 : 1996
Actual persons (full liability)	**15,725**	**24,989**	**27,834**	**28,989**	**58.9**	**11.4**	**4.1**
single enterprises	14,602	22,601	25,014	25,925	54.8	10.7	3.6
partnerships	760	1,897	2,291	2,541	149.6	20.8	10.9
limited partnerships	257	332	355	364	29.2	6.9	2.5
Legal entities (limited liability)	**2,749**	**2,824**	**2,894**	**2,942**	**2.7**	**2.5**	**1.7**
registered cooperatives	1,464	1,335	1,293	1,218	-8.8	-3.1	-5.8
limited companies	1,178	1,338	1,432	1,560	13.6	7.0	8.9
Total	**18,575**	**27,892**	**30,843**	**31,997**	**50.2**	**10.6**	**3.7**

* Enterprises with 1 ha or more farmland.
Source: BMELF.

Along with the restructuring of the LPGs, some former family farmers or their grown children tried to reestablish their holdings. In other cases, LPG members who did not have a family farm background tried to establish a new farm. More important, however, was the fact that the economic opportunities in East Germany attracted farmers from outside the New Laender. Initially many West German farmers tried to lease land near the former internal German border since the risk of these engagements was especially low. Later, West German or foreign farmers bought farms or entered into partnerships in all parts of East Germany.

Partnerships and single enterprises are fully liable and get preferential treatment in some areas such as taxation and social security (Forstner, 1995). Partnerships were often founded because of advantages in getting production quotas and public financial support and because the requirements to set up a large, modern farm were seen as too great for a sole entrepreneur. In partnerships, persons with different backgrounds with respect to financial resources and special knowledge often united in order to take advantage of their complementary resources, preferences, or skills. One example is the combination of a financially strong West German farmer and a farmer from East Germany who knows and is trusted by the local people.

The establishment of family farms, much desired by the federal government, was slow until 1991. After the legal basis for the asset distribution of LPGs was settled by the Agricultural Adjustment Act and promotion measures were improved, the establishment of family farms speeded up (Table 3).

The number of legal entities (including limited partnerships) has been stable over the seven years since 1992. However, within this group there was a shift towards limited companies while the number of cooperatives slightly declined. Most legal entities descend from a LPG, farm on a large scale, and have a great number of members. However, there are also single enterprises and partnerships which do not differ from legal entities in size and origin.

Figure 1:
Distribution of agricultural area among farms of different legal forms in the New Laender, 1992 to 1998 [1]

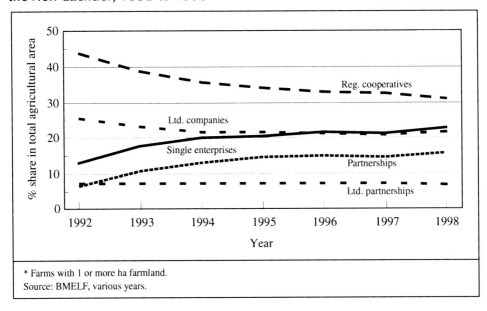

* Farms with 1 or more ha farmland.
Source: BMELF, various years.

Since large LPG-successor farms involve many persons with different interests, a full liability of all persons for each other's actions was not advisable. Considering that many problems (e.g. regulation of old debts, asset distribution, environmental damage) were not clarified when the decision for a new legal form had to be made, it is understandable that most LPGs chose a legal form with limited liability (registered cooperative or limited company). The German conversion law makes it possible that these holdings can further change their legal form in the future.

The share of farmland that belongs to the different legal forms corresponds to the numbers of farms of each organisaitonal type. Between 1992 and 1995, cooperatives in particular lost a major share of farmland, while single enterprises and partnerships gained considerable holdings (Figure 1).

By 1997 this tendency had lost momentum, but these organisations seem to be further increasing their market share. While in 1989 LPGs cultivated about 82% of the total farmland, the share of land farmed by LPG-successors has dropped to about 60% in 1998.

Table 4:
Agricultural holdings in the New Laender according to legal form, 1998[1]

Legal form	Holdings		Agricultural area		
	number	% of total	1,000 ha UAA	% of total	ha UAA per holding
Actual persons (full liability)	**28,989**	**90.6**	**2,555**	**45.6**	**88**
single enterprises	25,925	81.0	1,277	22.8	49
full time farms [2]	6,500	20.3	961	17.2	148
part time farms [2]	17,900	55.9	268	4.8	15
partnerships	2,541	7.9	889	15.9	350
limited partnerships	364	1.1	382	6.8	1,049
Legal entities (limited liability)	**3,008**	**9.4**	**3,046**	**54.4**	**1,013**
registered cooperatives	1,218	3.8	1,745	31.2	1,433
limited companies	1,560	4.9	1,207	21.5	774
Total	**31,997**	**100.0**	**5,601**	**100.0**	**175**

[1] Holdings with 1 ha or more farmland. [2] Data from 1997.

Source: BMELF; StBA.

The various legal forms differ both in size and production focus. On average, a cooperative is about ten times as large as a single (full-time) enterprise and about four times larger than a partnership. Limited companies are considerably smaller than cooperatives, but are marked by a high heterogeneity. Most single enterprises are very small and provide only part-time employment. An average full-time farm cultivates ten times the land of an average part-time farm (Table 4). To be sure, part-time farms often possess important social functions and in many cases render a significant contribution to the total household income, but from a production point of view they are of minor economic importance.

With the rising numbers and size of full-time single enterprises and partnerships, the dominance of large and very large farms has somewhat diminished (Figure 2). In 1991, farms with more than 1,000 ha farmland cultivated about 80% of the total farmland; seven years later, only about 55%. Farms with 200 to 1,000 ha farmland gained importance. Farms with less than 100 ha make up 73.5% of all farms, but cultivated only about 7% of total farmland (Table 5).

Figure 2:
Development of shares of different farm sizes in total agricultural area in the New Laender, 1991 to 1998

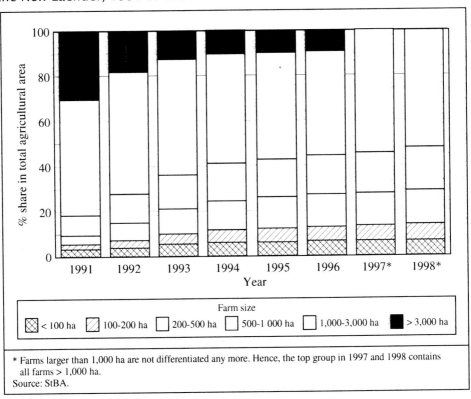

* Farms larger than 1,000 ha are not differentiated any more. Hence, the top group in 1997 and 1998 contains all farms > 1,000 ha.
Source: StBA.

Table 5 indicates how large the differences still are between the agricultural subsectors in East and West Germany. It is certainly not easy to design an agricultural policy that responds to the different preconditions in both parts of Germany.

While in West Germany small and medium-sized family farms dominate; in East Germany, large farms characterise agricultural production. Since ownership of farmland is vastly scattered in both East and West Germany, the differences in farm size lead to fairly high shares of leased land in East Germany. The share of leased land in East Germany amounts to 91% of farmland which is almost two times higher as in West Germany (48.1%).

Table 5:

Comparison of farm structure in the New and Old Laender, 1998

		New Laender	Old Laender
Farms	number	31,997	483,002
Agricultural area	1,000 ha	5,601	11,632
Ø Farmland per farm	ha	175	24
Farms < 20 ha	%	54.6	62.1
Farms > 100 ha	%	26.5	3.1
Farms > 1,000 ha	%	5.1	0.0
Farmland < 20 ha	%	1.8	18.2
Farmland > 100 ha	%	92.9	19.2
Farmland > 1,000 ha	%	52.0	0.2
Share of leased area	%	91.1	48.1*
Land rent	DM/ha/year	174.7	486.9
Land purchase price	DM/ha	6,365	33,629
Dairy cows	number	953	3,884
Farms > 100 dairy cows	%	38.6	0.6
Cows > 100 dairy cows	%	89.7	3.4
Milk quota	1,000 tons	6,242	21,525
Milk quota per ha	kg	1,114	1,850

* 1997.
Source: StBA; ZMP.

Further, in East Germany lease and purchase prices of agricultural land are comparatively low. Lease prices in West Germany average three times higher than in East Germany, purchase prices even five times higher. It is true that lease prices in East Germany have increased in recent years and have come nearer to the West German level, but presumably the much lower East German level will persist (Doll, et al., 1994; Isermeyer, 1995; Doll and Klare, 1998). Reasons for the rather low lease prices in East Germany are:

- Low livestock density and the low density of some highly-productive cultures (e.g. sugar beets, grapes, asparagus) result in comparatively low gross value added per hectare farmland.
- In the early 1990s, many farms were able to lease their farmland very cheaply on a long-term basis (10 to 12 years). In regions where the amount of leased state land is very high, the state agency responsible strongly influenced the level of land rents of other land owners.
- The scattered land ownership structure makes it rather difficult and costly for newcomers to collect the necessary information to compete successfully with the large-scale farm that currently dominates the local land market.

The differences between East and West Germany in purchase prices for agricultural land are mainly influenced by the opportunity to buy cheap (subsidised) state land. Beyond this, a low nonagricultural demand for farmland in East Germany and the financial restrictions on farms limit the increase of land prices (Doll and Klare, 1995).

Profitability of Farms

In the following section we present some results from a special analysis of the German farm accountancy network. We have included all full-time farms (actual persons and legal entities) from the sample of representative farms. The comparison of farms having different legal forms has some methodological problems. Therefore the following results must be considered with some caution.

Almost all key structural variables show great differences between East and West German farms. The average farm size of the East German farms in the sample amounts to 595 ha farmland compared to 47 ha in West Germany. However, the income capacity per labour unit is only slightly higher since East German farms have a much higher labour input and are less engaged in animal production. Special attention has to be paid to the vast difference in capital assets which for East Germany is only 26% of the

West German figure. The share of equity capital is considerably lower in East Germany which reflects the low share of own farmland and the high share of borrowed capital for financing investments.[8]

Table 6:

Comparison of agricultural holdings in East and West Germany, marketing year 1997/98[1]

		New Laender	Old Laender
Farms included in survey	number	1,419	7,737
Standard gross farm income	1,000 DM	618.4	72.7
	1,000 DM/WF unit	51.5	42.8
Size of farm	ha UAA	595	47
Share of lease land	%	46.7	45.2
Workforce (WF)	WF units	12.0	1.7
	WF units/100 ha	2.0	3.6
Livestock (LS)	LS units/100 ha	63.6	161.8
dairy cows	LS units/100 ha	23.0	45.2
pigs	LS units/100 ha	13.6	52.8
Assets	1,000 DM/ha	5.7	26.5
Capital assets	1,000 DM/WF unit	169.5	647.1
Share of equity capital	%	54.3	86.1
Direct subsidy payments	1,000 DM	433.2	26.7
	1,000 DM/100 ha	72.8	57.2
	1,000 DM/WF unit	36.1	15.7
Profit disposable to			
the entrepreneur (s)	1,000 DM/nWF unit[3]	74.9	39.3
Profit plus labour costs	1,000 DM/WF unit	44.5	38.0
Gross farm income	1,000 DM/WF unit	57.0	49.8
Return on equity[2]	%	2.6	-0.3

[1] Based on a representative farm accounting network of the BMELF; only full-time farms. [2] Equity includes the book value of land. [3] nWF=unpaid labour (family labour).

Source: BMELF, 1999.

The importance of direct subsidy payments is much higher in East Germany. This is due to the fact that East German farms are well equipped with land and often focus on cash crop production for which hectare-

---

[8] The share of equity capital of East German farms would even decrease if 'old' debts were included in their balance sheets.

related premia are paid. In contrast West German farms typically lack large amounts of farmland and focus on animal and special crop production for which no, or only minimal, hectare-related premia are paid.

On average, farms in the New Laender have reached a remarkable level of productivity and profitability. Average farm profit and gross farm income per labour unit are higher than in West Germany (Table 6). While farms in West Germany did not achieve a positive return on equity in the marketing year 1997/98, East German farms still realised 2.6%. The profit per unpaid family labour unit is nearly twice as high as in West Germany.

As was earlier shown, farm structure in East Germany is quite heterogeneous. Therefore it is necessary to further analyse the profitability and structure of farms according to legal form, type of farming, farm size, and success group. The results are presented in Tables 7 through 10 and can be summarised as follows:

- Partnerships are currently the most successful enterprises in East Germany. They achieved markedly more than 10% return on equity in each of the last three marketing years. They are very well equipped with land and show the highest labour productivity (standard gross farm income per workforce unit).
- Single enterprises currently realise a return on equity between 3 and 4%, compared to -0.3% in West Germany. The single enterprises in East Germany farm almost four times more farmland, but are equipped with much less asset capital than West German single enterprises. They achieve a higher labour productivity than their West German counterparts.
- On average in East Germany, legal entities are the least successful group, achieving a return on equity around zero. Their labour productivity is about on the West German level, well behind the levels achieved by partnerships and single enterprises in East Germany.
- Cash crop farms achieve the most favourable economic results, whatever their legal status. On the other side, forage-growing, single enterprises realise the least favourable results. Crop farms currently receive very high direct payments under the CAP. The payments per workforce unit are about double the amount for forage-growing farms. This leads to a serious dependency on future political decisions.

- While single enterprises and partnerships are predominantly specialised either in crop or animal farming, this is not true for legal entities. Since the reunion of specialised LPGs most successor farms own livestock in considerable quantities. Only a minor share of the legal entities are specialised in crop farming.
- Larger single enterprises realise better economic results than smaller ones. This holds true for both cash crop and forage-growing farms. Against that, profitability and factor productivity within the group of legal entities do not differ according to farm size.
- An analysis of farms in Mecklenburg-Pommerania (in the northern part of East Germany) showed that 40% of forage-growing farms and 31% of cash crop farms were unable to realise a positive return on investments in the marketing year 1997/98 (Ministry for Agriculture in Mecklenburg-Pommerania, 1999).

In the early 1990s, the reform of the CAP brought about the stepwise reduction of intervention prices for major cash crops and relatively high, hectare-related compensation payments. This framework delivered favourable and reliable conditions for large cash crop farms, at least for the moment. As farms in East Germany lacked major financial resources and were keen to limit risk, crop production had some advantages over animal farming. Crop production required much less capital than the establishment of milk or pig production. The low land rents were also a favourable precondition for large cash crop farms in East Germany, too. The outstanding economic performance of the East German partnerships must be seen against this background. The legal entities, mostly successors of the LPGs, found it much more difficult to convert their farms to pure crop farms. Cooperatives in particular had to provide employment for their working members. Laying off more members would have been very dangerous, because these members might have decided to withdraw their capital and their land from of the cooperative. Also many cooperative leaders regarded it as their social obligation to minimise labour cuts. German tax codes, as well as the „old debt" regulations, support the approach of the legal entitities in not showing profits, because only a small portion of such profit would stay with the farm.

Table 7:
Comparison of agricultural holdings in the New and Old Laender according to legal form, marketing year 1997/98[1]

		New Laender				Old Laender		
		Single enterprises[2]	Partnerships	Legal entities	Total	Single enterprises[2]	Partnerships	Total
Farms included in survey	number	855	205	358	1,419	7,278	459	7,737
Standard gross farm income	1,000 DM	135.5	387.0	1,381.6	618.4	71.6	109.3	72.7
	1,000 DM/WF unit	56.4	74.4	47.5	51.5	42.6	48.0	42.8
Size of farm	ha UAA	168	400	1,524	595	46	66	47
Share of leased land	%	87.2	95.6	96.6	95.2	56.0	74.9	56.8
Workforce (WF)	WF units	2.4	5.2	35.4	12.0	1.7	2.3	1.7
	WF units/100 ha	1.4	1.3	2.3	2.0	3.6	3.5	3.6
Livestock (LS)	LS units/100 ha	37.0	34.1	76.3	63.6	162.1	155.5	161.8
dairy cows	LS units/100 ha	11.9	18.0	26.4	23.0	45.4	39.8	45.2
pigs	LS units/100 ha	4.8	2.1	18.3	13.6	52.4	62.2	52.8
Assets	1,000 DM/ha	5.2	4.2	6.2	5.7	26.8	19.9	26.5
Capital assets	1,000 DM/WF unit	277.5	221.6	149.1	169.6	653.8	486.6	647.1
Share of equity capital	%	53.7	37.1	57.4	54.3	86.2	81.9	86.1
Direct subsidy payments	1,000 DM	122.5	283.2	1,118.4	433.2	26.4	37.2	26.7
	1,000 DM/100 ha	72.7	70.7	73.4	72.8	57.2	56.9	57.2
	1,000 DM/WF unit	51.0	54.4	31.6	36.1	15.7	16.4	15.7
Profit disposable to the entrepreneur	1,000 DM/nWF unit[4]	48.5	86.2	-	-	36.6	42.1	-
Profit plus labour costs	1,000 DM/WF unit	46.7	62.0	41.7	44.5	37.8	40.3	37.9
Return on equity[3]	%	4.2	17.9	0.6	2.6	-0.3	0.4	-0.3

[1] Based on a representative farm accounting network of the BMELF. [2] Only full-time farms. [3] Equity includes the book value of land.
[4] nWF=unpaid labour (family labour)

Source: BMELF, 1999.

Hence the great difference in the performance figures between legal entities on the one hand, and single farms and partnerships on the other should be analysed very carefully. It can be expected, that partnerships will continue to have better figures in the future, because they manage farm sizes which enable them to take advantage of economies of size without the drawbacks of very large farms (transaction costs, etc.). However, it can also be expected that the backwardness of legal entities will diminish.

Table 8:

Comparison of farms in the New Laender, by legal status and type of farming, marketing year 1997/98[1]

		Single enterprises[2]		Partnerships		Legal entities	
		Cash crop	Forage growing	Cash crop	Forage growing	Cash crop	Forage growing
Standard gross farm income	1,000 DM	169.5	93.7	513.5	261.0	1,543.4	1,778.7
	1,000 DM/WF unit	68.8	41.1	88.0	57.0	51.2	45.2
Size of farm	ha UAA	219	111	577	234	1,715	1,453
Share of lease land	%	87.3	86.6	96.0	94.6	96.7	97.3
Workforce (WF)	WF units	2.5	2.3	5.8	4.6	30.1	39.4
	WF units/100 ha	1.1	2.1	1.0	2.0	1.8	2.7
Livestock (LS)	LS units/100 ha	12.9	85.3	11.4	87.6	38.7	89.6
Direct subsidy payments	1,000 DM	162.8	74.7	411.0	162.2	1,216.0	1,077.5
	1,000 DM/100 ha	74.3	67.6	71.3	69.3	70.9	74.2
	1,000 DM/WF unit	66.0	32.8	70.4	35.4	40.4	27.4
Profit plus labour costs	1,000 DM/WF unit	57.4	33.9	76.0	45.6	46.0	38.8
Gross farm income	1,000 DM/WF unit	50.1	27.5	71.1	42.3	39.3	33.7
Return on equity[3]	%	7.5	-0.6	22.3	10.1	2.1	-0.7

[1] Based on a representative farm accounting network of the Federal Ministry for Agriculture. [2] Only full-time farms. [3] Equity includes the book value of land.

Source: BMELF, 1999.

Table 9:

Comparison of farms in the New Laender, by farm size and type of farming, marketing year 1997/98 [1]

Legal form		Single Enterprises [2]			Legal Entities		
Standard gross farm income in 1,000 DM		< 100	100 to <200	200 to <500	500 to <1.000	1.000 to < 2.000	2.000 to 3.000
		Cash crop farms					
Standard gross farm income	1,000 DM/WF unit	33.1	66.9	87.4	51.5	50.8	53.1
Workforce (WF)	WF units/100 ha	1.6	1.1	1.0	1.4	1.7	1.9
Livestock (LS)	LS units/100 ha	27.7	12.6	8.1	30.2	36.5	45.3
Profit plus labour costs	1,000 DM/WF unit	30.4	53.0	72.0	46.6	45.8	46.6
Gross farm income	1,000 DM/WF unit	42.1	74.9	102.7	63.5	60.8	61.4
Return on equity [3]	%	-3.2	5.3	12.8	2.3	2.7	2.0
		Forage-growing farms					
Standard gross farm income	1,000 DM/WF unit	27.8	53.8	62.4	46.6	45.1	43.6
Workforce (WF)	WF units/100 ha	2.4	1.9	1.6	2.2	2.5	2.7
Livestock (LS)	LS units/100 ha	91.2	80.8	81.1	73.6	86.6	86.3
Profit plus labour costs	1,000 DM/WF unit	26.6	38.9	52.6	38.0	38.2	39.6
Gross farm income	1,000 DM/WF unit	33.5	53.4	69.2	49.5	46.3	45.7
Return on equity [3]	%	-4.1	1.8	12.0	-1.8	-1.2	-0.2

[1] Farm size according to Standard Gross Income; based on a representative farm accounting network of the Federal Ministry for Agriculture. [2] Only full-time farms. [3] Equity includes the book value of land.

Source: BMELF, 1999.

Table 10:
Comparison of profitability of agricultural holdings in the New Laender (profit plus labour costs per worker)[1]

		Single enterprises [2]		Partnerships		Legal entities	
		25 % least profitable	25 % most profitable	25 % least profitable	25 % most profitable	25 % least profitable	25 % most profitable
Standard gross farm income	1,000 DM	69.3	208.2	254.3	485.7	1,637.4	1,474.6
	1,000 DM/WF unit	34.8	82.3	48.7	111.4	40.4	59.5
Workforce (WF)	WF units	2.0	2.5	5.2	4.4	40.5	24.8
	WF units/100 ha	2.0	1.1	1.6	1.0	2.7	1.9
Livestock (LS)	LS units/100 ha	45.0	37.1	39.0	24.8	78.1	78.8
Capital assets	1,000 DM/WF unit	260.9	335.5	189.0	279.1	130.0	188.4
Share of equity capital	%	56.8	52.8	29.8	44.9	59.6	56.3
Direct subsidy payments	1,000 DM	66.3	168.6	212.0	315.2	1,105.0	960.7
	1,000 DM/WF unit	33.3	66.7	40.6	72.3	27.3	38.7
Profit plus labour costs	1,000 DM/WF unit	9.8	85.6	21.8	111.2	28.3	61.2
Gross farm income	1,000 DM/WF unit	21.5	111.0	37.2	142.8	36.2	75.9
Return on equity [3]	%	-14.1	18.2	-16.3	35.0	-4.6	5.9

[1] Based on a representative farm accounting network of the Federal Ministry for Agriculture. [2] Only full-time farms. [3] Equity includes the book value of land.

Source: BMELF, 1999.

Assessment and Outlook

In the early 1990s, agricultural economists and politicians vehemently disputed the future development of the agricultural sector in the New Laender. The opinions were fairly conflicting and a blueprint for a comparable transformation process did not exist. The central question was whether to build up an agricultural structure based on family farms as in the Old Laender, or to leave the structural development mainly to the market forces, i.e. to give all organisational forms of enterprises their own chances. The basic decision was made in favour of the market solution. The Agricultural Adjustment Law laid the foundation, and various support programmes were aimed at preventing discrimination toward any particular form of farming. The transformation process and the now visible results give the following answers to this unique empirical experiment (Isermeyer, 1995):

- The basic political decision in favour of competition in a market framework now seems to have been correct. Considering all of the facts, the transformation process has largely succeeded. But it must also be remembered that huge amounts of taxpayer money have been invested in this sector and that, despite this investment, many farms, both large and small, are now in financial straits.
- Experience has shown that large farms can be managed efficiently and successfully. Large farms are well suited for the use of new technological and organisational solutions (e.g. precision farming). Equipped with modern technology, even very large farms no longer need hundreds of workers as they did some decades ago. Hence, the importance of transaction costs should not be overestimated. If management is sound, even very large farms seem to be competitive; if the management is faulty, it should be replaced.
- The widely advocated standpoint that smaller farms are more favourable for the environment than larger farms can not be supported, at least not in such a generalised way. In both East and West Germany, there are well-managed and poorly-managed farms in all size groups. The future will show which farms are best suited to respond to the increasing requirements of environmental protection laws.
- The developments in the New Laender cleared some ideological barriers from the minds of many West German farmers and politicians. More and more partnerships are being established, and more importance is being laid on accounting results and total cost calculations rather than solely on gross margin calculation. This will further contribute to structural adjustment. Another interesting experience that could also be beneficial for West Germany, is the organisation of the lease-land market which, despite scattered ownership, continously leads to the large land plots that are necessary for efficient farming. In the longer run, the advantages of large plots will also force West German farmers to cooperate and apply plot exchange mechanisms similar to those currently established in the East.
- In East Germany, structural adjustments have slowed considerably. Meanwhile, the number of farms has increased only slightly. The number of large farms and their share in total farmland remains

constant. However, there are many farms suffering from serious economic problems. This applies to farms of different types and sizes. It can be expected that a significant number of farm enterprises will have to stop farming. This will presumably not greatly change farm size structure since the respective production units will largely be taken over by other large and competitive farms.

- On average, crop farming turned out to be more profitable than animal farming. Many new and reestablished farms focused on cash crop production first, since the necessary capital and other requirements for start-up were lower than with animal production. The guaranteed prices and land-related direct payments which were introduced as a result of the CAP reform in 1992 offered an environment well-suited to the crop farming. It must be kept in mind, however, that crop farms are highly dependent on subsidies that may not be paid forever, and that increasing land rents will steadily reduce profits.

- The drastic reduction in animal production was no temporary event. Pig production, which was hardest hit, is now being fostered to regain production shares. However, the success of such attempts has been very limited. Currently only a few farmers, many from the Netherlands, are trying to establish pig production units in the New Laender. A significant rise in pig production could probably be expected only if strict environmental constraints forced the current production centres to drastically reduce pig numbers. For the New Laender, it is obviously very difficult to catch up the economies of agglomeration and the special know-how that has accumulated in other regions of West Germany and the EU over decades.

- Problems in LPG-successor farms arise from the large number of members and their divergent interests. In the past, agricultural cooperatives seemed to suffer more than other farm types from inefficient management structures. However, the process of restructuring LPG-successor farms has changed their forms. Now, there is a clear tendency in many cooperatives to concentrate the farm capital on a small number of members (or shareholders) who are prepared to take responsibility and liability for the farm.

- Agenda 2000 provides a clear political framework for the next seven years. For the subsequent period, the upcoming WTO round will probably lead to a further decline of export subsidies, import barriers, and internal support. International comparisons of costs of production show that at least the crop farms in East Germany seem to be rather well prepared for intensified international competition. To what extent livestock farms in East Germany can further reduce their costs of production in order to improve their international competitiveness must be analysed more carefully in the near future.

References

Agricultural Adjustment Law (Landwirtschaftsanpassungsgesetz) 3 July 1991. BGBl. I P. 2082.

Bell, W. (1992). Enteignungen in der Landwirtschaft der DDR nach 1949 und deren politische Hintergründe. (Expropriations in Agriculture of the GDR after 1949 and their Political Background.)

BMELF (Various Years) (Federal Ministry for Agriculture). Agricultural Report of the Government.

BMELF (1991) (Federal Ministry for Agriculture). Agrarbericht der Bundesregierung 1991. (Agricultural Report of the Government 1991.) P. 138-152.

BMELF (1995) (Federal Ministry for Agriculture). Agrarbericht der Bundesregierung 1995, Materialband. (Agricultural Report of the Government 1995, Material Collection.) P. 25-27.

BMELF (1996) (Federal Ministry for Agriculture). Agrarwirtschaft in den neuen Ländern - Aktuelle Situation und Maßnahmen. (Agriculture in the New Laender - Current Situation and Measures.) P. 24-25.

BMELF (1998) (Federal Ministry for Agriculture). Agrarbericht der Bundesregierung 1995. (Agricultural Report of the Government 1998.) P. 84.

BMELF (1999) (Federal Ministry for Agriculture). Results of a Special Analysis of the Representative Farm Accounting Network (Unpublished Data).

Doll, H. and Klare, K. (1998). Land bleibt teuer. (Land Remains Expensive.) DLG-Mitteilungen, No. 8, P. 12-16.

Doll, H. and Klare, K. (1995). Empirische Analyse der regionalen landwirtschaftlichen Bodenmärkte in den neuen Bundesländern. (Empirical Analysis of Regional

Markets for Farmland in the New Laender.) Landbauforschung Völkenrode, No. 4, P. 205 - 217.

Doll, H., Günther, H.-J. and Klare, K. (1994). Empirische Analyse der Pachtmärkte in Mecklenburg-Vorpommern. (Empirical Analysis of Land Lease Markets in Mecklenburg-Pommerania.) Landbauforschung Völkenrode, No. 1, P. 54-66.

Eberhard, M. (1991). Betriebsorganisation der LPG bei veränderten Markt- und Preisbedingungen. (Organisation of LPG under a Changed Market and Price Environment.) In: Schinke, E. and Merl, S. (Eds.), Agrarwirtschaft und Agrarpolitik in der ehemaligen DDR im Umbruch. (Agriculture and Agricultural Policy in Former GDR in Transition). P. 133 - 150.

Forstner, B. (1995). Schlechte Karten für juristische Personen. (Poor Chances for Legal Entities.) DLG-Mitteilungen, No. 6, P. 46-49.

Isermeyer, F. (1991). Umstrukturierung der Landwirtschaft in den neuen Bundesländern - Zwischenbilanz nach einem Jahr deutscher Einheit (Restructuring East Germany´s Agriculture - Interim Balance after One Year of German Reunification.) Agrarwirtschaft 40, P. 294 - 305.

Isermeyer, F. (1995). Lehren aus der Umstrukturierung der ostdeutschen Landwirtschaft für die Weiterentwicklung in den westdeutschen Ländern. (Lessons from the Transformation Process of East German Agriculture for the Further Development in the West German Laender.) Institute of Farm Economics of the Federal Agricultural Research Centre. Working Paper No. 1/95.

König, W. and Isermeyer, F. (1996). Eine empirische Untersuchung des Anpassungsverhaltens landwirtschaftlicher Unternehmen im Übergang zur Marktwirtschaft. (Empirical Analysis of the Adjustment of Farms during the Transition to Market Economy.) In: Kirschke, D., Odening, M. and Schade, G. (Eds.), Agrarstrukturentwicklungen und Agrarpolitik. (Agricultural Developments and Policy.) Schriften der Gesellschaft für Wirtschafts- und Sozialwissenschaften des Landbaues. Volume 32. P. 335-346.

Ministerium für Ernährung, Landwirtschaft, Forsten und Fischerei Mecklenburg-Vorpommern (1999) (Ministry for Agriculture in Mecklenburg-Pommerania). Agrarbericht 1999 des Landes Mecklenburg-Vorpommern. (Agricultural Report 1999). P. 30-35.

Reichelt, H. (1992). Die Landwirtschaft in der ehemaligen DDR - Probleme, Erkenntnisse, Entwicklungen. (Agriculture in the GDR - Problems, Perceptions, Developments). Berichte über Landwirtschaft 70, P. 117-136.

Schmitt, G. (1993). Strukturanpassung in den neuen Bundesländern. (Structural Adjustment in the New Laender.) Agra Europe 46/93, Sonderbeilage. (Special Supplement.)

Schweizer, D. (1994). Das Recht der landwirtschaftlichen Betriebe nach dem Landwirtschaftsanpassungsgesetz. (The law of agricultural holdings according to the Agricultural Adjustment Law.) 2. Edition.

StBA (Various Years) (Federal Agency for Statistics). Fachserie 3, verschiedene Reihen. (Subject Series 3, Various Topics.)

Thiele, H. (1998). Dekollektivierung und Umstrukturierung des Agrarsektors der neuen Bundesländer - Eine gesamtwirtschaftliche und sektorale Analyse von Politikmaßnahman. (De-collectivisation and Restructuring of the Agricultural Sector in the New Länder - A Macro Economic and Sectoral Analysis of Policy Measures.)

Weber, A. (1991). Zur Agrarpolitik in der ehemaligen SBZ/DDR - Rückblick und Ausblick. (Aspects to the Agricultural Policy in the Former GDR - Retrospective and Outlook) In: Merl, S. and Schinke, E. (Eds.), Agrarwirtschaft und Agrarpolitik in der ehemaligen DDR im Umbruch. (Agriculture and Agricultural Policy in Former GDR in Transition.) P. 53-70.

Welschof, J. (1995). Die strukturelle und institutionelle Transformation der landwirtschaftlichen Unternehmen in den neuen Bundesländern. (Structural and Institutional Transformation of Agricultural Enterprises in the New Laender.) Studien zur Wirtschafts- und Agrarpolitik, Volume 14.

Wolffram, R.; Bongaerts, R; Simonis, J. (1996). Überlegungen zur Korrektur von Fehlentwicklungen in der Schlachthofstruktur der neuen Bundesländer. (Reflections on Correction of Development Mistakes in the Slaughterhouse Structure in East Germany.) Agrarwirtschaft 45, P. 335-342.

ZMP (Various Years)(Central Agency for Market and Price Reports on Agricultural Products). ZMP-Bilanz Getreide, Ölsaaten, Futtermittel. (ZMP Balance Cereals, Oilseeds, Feedstuffs.)

ZMP (Various Years)(Central Agency for Market and Price Reports on Agricultural Products). ZMP-Bilanz Milch. (ZMP Balance Milk.)

ZMP (Various Years)(Central Agency for Market and Price Reports on Agricultural Products). ZMP-Bilanz Vieh und Fleisch. (ZMP Balance Animals and Meat.)

ZMP (1999)(Central Agency for Market and Price Reports on Agricultural Products). ZMP-Bilanz Milch 99. (ZMP Balance Milk 99.)

Chapter 5

Macroeconomic Framework and Implications for German Agriculture after Reunification*

Dieter Kirschke and Steffen Noleppa
Institute of Economic and Social Sciences of Agriculture,
Humboldt University of Berlin

Introduction

When Hungary opened its Western border to citizens of the German Democratic Republic (GDR) in September 1989, it was a decisive step towards the reunification of Germany that no one expected to occur only one year later. A few weeks later, on November 9, 1989, the Berlin Wall broke down. On July 1, 1990 both German states established a currency union, and on October 3, 1990 Germany celebrated its reunification.

The reunification of Germany has been both a political and economic challenge for East and West Germany. Four decades of communist regime had virtually ruined the East German economy; thus, the reunification of Germany imposed severe structural adjustment requirements for the East German economy, but also for the Western part of the country. In this chapter the dimension of this adjustment process will be described. The particular development on the labour market is examined, and the financial consequences of the reunification are discussed. The macroeconomic framework of the German reunification has had a particular importance for the agricultural sector. The agricultural sector itself has played a genuine role in this adjustment process, and the

---

* The authors would like to thank Michael C. Burda, Stefan Tangermann, and Harald von Witzke for valuable support and comments.

implications for the sector of the macroeconomic framework and adjustment process have been significant.

The Dimension of Structural Adjustment

It has become popular to describe structural adjustment in the transition process with a U-curve or a J-curve (Blanchard, 1996; Siebert, 1992). In all transition countries this process has started with a severe cut in production, and this was also true for East Germany. Industrial production in East Germany dropped by 40 percent after the establishment of the currency union and in 1991 amounted to only one-third of the level in the former GDR. The gross national product (GNP), which is less reactive than industrial production, dropped by 35 percent in the same period (Sinn and Sinn, 1994). In 1991 alone, the gross domestic product (GDP) in East Germany dropped by 19 percent (Deutsche Bundesbank, 1998). By the end of that same year unemployment had increased to 30 percent (Sinn and Sinn, 1994).

Without question the East German economy had to be totally restructured after reunification. In 1989, the former GDR was practically bankrupt. The communist regime led to enormous structural distortions due to wrong incentives and institutions, lack of know-how, and prices that had nothing to do with international competitiveness (Sinn and Sinn, 1994). Table 1 shows the structural differences between the East and West German economies before reunification. Labour shares show that the former GDR had allocated considerably more resources to agriculture, energy and mining, and the processing industry compared to West Germany. On the other hand labour shares for trade and services and government were much lower in the former GDR. The same picture can be drawn from a comparison of the production shares. The figures particularly underline deficits in the services sector in the GDR economy. Table 1 reveals severe adjustment needs of the East German economy after reunification, mostly due to the shift of a planned to a market economy, but certainly also due to the relatively low income level as compared to the West.

The production structure in East Germany quickly became similar to that of West Germany, but the macroeconomic adjustment process has faced much friction and is still, ten years after reunification, far from completion. Figure 1 shows the development of selected macroeconomic indicators in East Germany compared to West Germany.

Table 1:
Structural differences between the former German Democratic Republic (GDR) and West Germany, end of the 1980s

	GDR	West Germany
Labour shares of various sectors (percent)		
Agriculture	9.9	4.2
Energy and mining	3.2	1.4
Processing industry	34.1	29.7
Construction	6.1	6.6
Trade	7.8	14.5
Transport	6.8	5.6
Services and government	32.4	38.0
Production shares of various sectors (percent)		
Agriculture	9.0	2.0
Processing industry and mining	59.0	41.0
Trade and transport	12.0	15.0
Services	7.0	29.0
Government	12.0	13.0

Source: Koester and Brooks (1998) and Sinn and Sinn (1994)

The catch-up process was considerable during the first years following reunification, but has somewhat stagnated since. Today workers in East Germany still have a gross wage level 25 percent below the West German level. In 1998 the GNP per capita in East Germany amounted to only 56 percent of the West German level. Figure 1 also shows that the investment level per capita in East Germany has surpassed that in West Germany, particularly for investments in building. It should be noted, however, that in recent years the investment level in East Germany has stagnated and today seems in line with the West German level. This is certainly not sufficient for the East to truly catch up, and there is widespread concern about this development. In fact, Sinn and Sinn (1994) argue that investment requirements for the East German economy would amount to as much as 1 000 billion Deutsche marks (DM) using West German capital intensity as a reference. This figure may be questioned, but there is certainly a need for much higher investment rates in East Germany than the ones that were observed during the adjustment process. Such trends have led to some disillusionment about the catch-up process of East Germany after reunification, though a more optimistic view is advocated in view of the enormous achievements so far (Burda and Funke, 1995).

Figure 1:
Development of selected macroeconomic indicators in East Germany (West Germany = 100 percent)

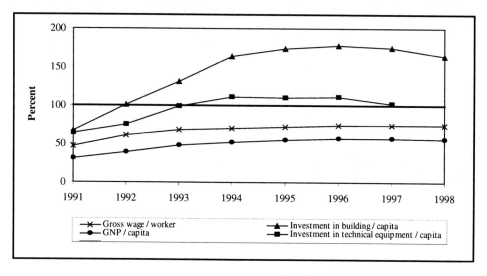

Source: DIW, IfW, and IWH (1999)

There has been continuous debate on the economic adjustment process in East Germany, and there is a rising criticism on the policies that have been pursued. It is evident that the currency union marked a drastic starting-point for the adjustment of the East German economy which at first was actually a breakdown. Most experts, however, would agree that in view of potential migration problems there was no real alternative.[1] The chosen exchange rates of 1 Mark (M) to 1 DM for private assets (up to a limited amount), 2 M to 1 DM for other assets and liabilities (1.80 M to 1 DM effectively), and a 1 M to 1 DM conversion for prices and existing

---

[1] Interestingly, intra-German migration was restricted due to several factors (Burda, 1993).

contracts were criticised at first, but in today's view seem acceptable (Sinn and Sinn, 1994; Tietmeyer, 1999).

Figure 2:
Prices for agricultural products in the former GDR (percent of average producer prices in West Germany) a

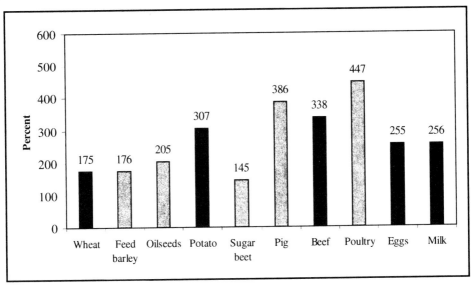

a Exchange rate: 1 M = 1 DM
Source: Koester and Brooks (1998)

Hence, the real policy problem has been the design of a policy framework for the adjustment process. Two factors are considered to have had a major negative impact on the catch-up process in East Germany: the privatisation policy and the wage policy. Germany has emphasised restitution instead of compensation as a privatisation principle, and this has severely hampered the reorganisation of firms and investment decisions. Relatively high wages in view of the low productivity, have aggravated the problem. As a consequence, large public transfers from West to East have been necessary and have, been enacted to increase the East German consumption level. Today, in East Germany the unique situation exists in which the absorption of goods and services is about 50 percent higher than production (Sinn, 1999).

The figures in table 1 also point to the particular adjustment needs of the agricultural sector in East Germany after reunification. As for all other sectors, incentive and price distortions for this sector were enormous in the former GDR. Figure 2 shows the prices of agricultural products in the GDR compared to average producer prices in West Germany. The absolute price level of agricultural products in the former GDR was much higher than in West Germany, valued at the exchange rate of 1 M to 1 DM effective with the establishment of the currency union in 1990. Even more interesting is the fact that relative prices in the former GDR were considerably distorted in favour of animal production. Hence, adjustment needs after reunification have been enormous with respect to prices of agricultural products. The adjustment problem is aggravated due to the fact that factor prices were relatively low in the former GDR and tended to increase after reunification.[2]

Häger, Kirschke, and Noleppa (1999) have analysed structural adjustments in East German agriculture in more detail. They distinguish between behaviour, price, and technology effects on agricultural production during the adjustment process. The behaviour effect describes the production change resulting from the shift from cost coverage to profit maximisation. This effect was typically negative, in the range of 10 to 40 percent as compared to the original production level. The price effect also was negative in the range of 10 to 30 percent. The technology effect, on the other hand, was positive, in the range of 30 to 100 percent. This underlines the specific importance of technology and knowledge transfers in the transition process and the potential impact that this might have on production.

The macroeconomic adjustment process in the East German economy has had specific implications for the agricultural sector. First, agriculture is a major sector in the adjustment process itself. The sector has contributed considerably to the overall economic picture that we observe today, e.g. in terms of unemployment. Additionally the sector has benefited and suffered from the same policy framework as the economy as a whole. Second, agriculture in East Germany continues to operate in a rather weak macroeconomic environment. This is of particular relevance when judging the opportunity costs of agricultural production and labour. Third,

---

[2] This is certainly true for labour costs and technical equipment, e.g., but not for fertiliser and pesticides (HÄGER, 1999).

economic adjustment not only causes problems, but offers opportunities. Despite further adjustment needs, today the most efficient farms in Germany can be found in the East, and the average labour income in agriculture is higher in the East than in the West (BMELF, 1999). This is mostly due to radical adjustments on the labour market, i.e. a massive shedding of labour from East German agriculture, whereas in West Germany adjustments in the agricultural labour force still need to be made.

Development of the Labour Market

Structural adjustment in transition economies affects all resources, but a most important aspect of the transition process is unemployment and the need to create new jobs and develop human capital. Following reunification unemployment in East Germany was, and continues to be, rife. The 'official' unemployment rate for East Germany has been around 15 percent reaching 18 percent in 1997 (Deutsche Bundesbank, 1998). It must be recognised, though, that these figures greatly underestimate the problem. First, the labour force in East Germany has decreased considerably from the former GDR level due to migration (Burda, 1993), employment opportunities in West Germany, and a reduction in the labour supply (e.g. many women decided to withdraw from the labour market). Second, enormous government subsidies on the labour market have helped to keep the 'official' unemployment rate artificially low. It is estimated that without such interventions the 'effective' unemployment rate would have reached 30 percent, leaving one person out of three without a job. Such a depression is without parallel in modern economic history. Even the world economic crisis from 1928 to 1933 did not have such drastic consequences, since it took place over a longer period of time and resulted in smaller production cuts (Sinn and Sinn, 1994).

Figure 3 shows the development of employment in the East German economy since reunification. Total employment opportunities, excluding government subsidies, dropped to about 65 percent of the former GDR level in only three years and have remained in the range of 60 to 65 percent since. The labour force has dropped by more than 2 million people; half due to migration and a reduction in the labour supply, half due to early retirement and East Germans being employed in West

Germany. Figure 3 also reveals that there are no signs of recovery and that the labour market seems to have stagnated at the described low level. In fact, employment expectations for the East German economy remain rather negative and are considerably worse compared to West Germany (IAB, 1998). The overall picture must be differentiated by sectors. In several sectors, like agriculture or energy and mining, job opportunities have continuously and drastically declined. In the construction sector there was a boom in the first years after reunification, but now employment has declined in this sector, too. On the other hand, employment opportunities have been continuously increasing in the services sector (IAB, 1998).

Figure 3:
Labour market developments in East Germany, 1989-1997

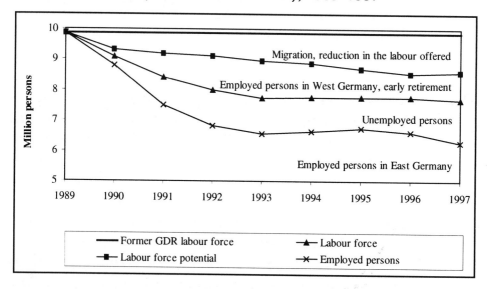

Source: IAB (1998)

In the agricultural sector of East Germany the drop in employment opportunities has been particularly drastic. Three years after reunification the employment level in agriculture dropped to 35 percent of the former GDR level. This drop in agricultural employment was particularly high in Brandenburg and Saxony-Anhalt. As a general rule, by 1992 only half of workers who left the agricultural sector found new jobs in other sectors, and the other half remained unemployed (Häger and Hagelschuer, 1996). Hence, the agricultural sector significantly contributed to high East

Macroeconomic Framework after Reunification

German unemployment still observed today. Of the 915 000 persons who worked in the former GDR agricultural sector, only 145 000 were employed in that sector in 1998 (BMELF, 1999).

The development of the wage level in East Germany is a point of particular interest. Figure 4 shows selected indicators for the East German labour market after reunification, compared to West Germany. East German labour productivity increased quickly after reunification from only 30 percent of West German labour productivity, but has stagnated at around 60 percent in recent years. Gross income per employee has always been above the productivity level; it has increased from about 45 percent to about 75 percent of the West German level (Pohl and Ragnitz, 1998). The obvious problem is that East German workers still earn less than their West German colleagues, but that wage unit costs have been and still are considerably higher in East Germany. This is considered, as stated above, one of the major causes for the insufficient adjustment of the East German economy. The problem becomes apparent in the development of the unemployment rate in East Germany, which continues to be much higher compared to West Germany.

Figure 4:
Development of selected labour market indicators in East Germany
(West Germany = 100 percent)

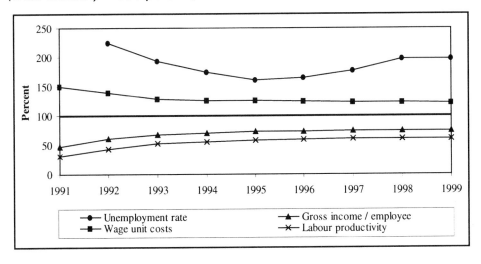

Source: ifo (1999), Pohl and Ragnitz (1998), StBA (various issues)

The East German labour market problem can be further visualised by a transformation path as presented by Sinn (1999). Figure 5 shows wage rates, as a percentage of the West German level, and employment levels in East Germany after reunification in percent of the 1989 level. The figure shows that employment has gone down but wages have increased from 1989 to 1992. Since then, both wages and employment levels have moved slightly upwards. Basically, the figure shows the stagnation on the East German labour market and in the adjustment process in recent years.

Figure 5:
The transformation path in East Germany

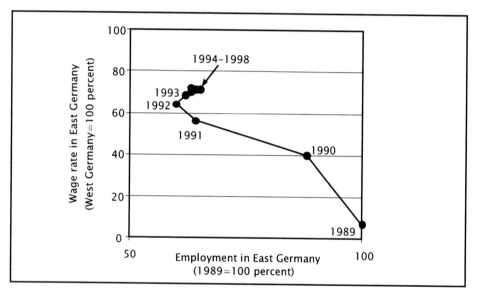

Source: Sinn (1999)

What are the implications of developments on the East German labour market for the agricultural sector? First, the agricultural sector has considerably contributed to the overall labour market problem, and was certainly not a sector that absorbed unemployed force in the transition process, contrary to the situation in many Central and Eastern European countries, where agriculture served as a buffer for unemployment. Despite the drastic decline of labour employed in the sector there seem to be prospects for further adjustment. Calculations have shown that the labour force in the sector still would need to be reduced by about 25 percent if farms were to fully adjust to the new economic framework (Kirschke et al.,

1998). Second, developments in the labour market in East Germany have led to particularly severe problems in rural areas. Unemployment in these areas is typically higher than in urban and industrialised areas. Since agriculture has played a relatively more important role in the north than in the south, the transition process has tended to accentuate adjustment problems in rural areas in Northern East Germany, which also have a relatively weak industrial base. In view of such developments, there has been notable debate in East Germany on the contribution of agricultural policy to employment in rural areas. Taking into account the scale of the sector on the one hand, and the dimension of the labour market problem on the other, there should be no illusions on the part of agricultural policy makers: agricultural policy will not solve labour market problems.

Financing the German Reunification

The German reunification has had strong financial implications for Germany as a whole and in particular for the public sector. For this sector the reunification resulted in immense transfers from West to East. These challenges had to be met in a framework of a sound, yet weakening West German economy. Rigidities on markets, especially the labour market, and distorting government interventions were criticised long before reunification. From 1960 to 1990 government expenditure increased from about one-third to 46.1 percent of GDP and government revenues from about 30 to 44.0 percent of GDP. As a consequence the budget deficit rose to 2.1 percent of GDP by 1990 and public debt amounted to 45.6 percent of GDP compared to 17.4 percent in 1960 (ifo, 1999).

The economic situation worsened under and due to reunification. By 1995 government expenditures had risen to 50.9 percent of GDP and receipts to 47.6 percent. The budget deficit increased to 3.3 percent in 1995 and 1996, slightly decreasing since, whereas public debt has increased continuously and is expected to climb to 61.5 percent of GDP in 1999 (ifo, 1999). It is difficult to assess how much of this has really been due to German reunification or is merely the consequence of neglected economic reforms. In recent years the integration of the European Union (EU) and the introduction of the euro has undoubtedly put pressure on Germany and has helped bring readjustment of the overall economy.

Figure 6:
The reversal of the German current account after reunification

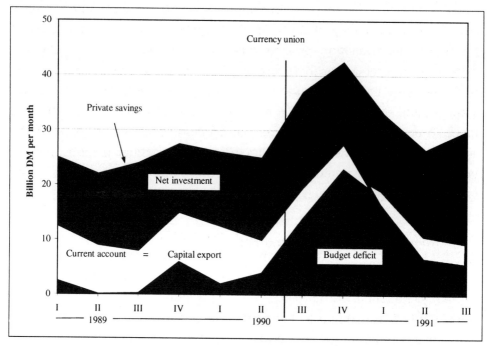

Source: Sinn and Sinn (1994)

With respect to the balance of payment, the German current account has traditionally been positive, allowing for considerable capital exports. This situation was reversed in 1991. The reunification resulted in increased capital demand and a decline in the trade balance due to increased imports and decreased exports; as a consequence, the current account became negative after reunification and Germany no longer exported, but imported, capital. The reversal of the balance of payment situation is illustrated by figure 6. The typical situation before 1990 was that private savings in West Germany were used to finance private investments, capital export, and a small budget deficit. After reunification private savings increased and, together with capital imports, had to finance increasing private investments and an increasing budget deficit. The German current account was negative, although decreasing up to 1998. Throughout this period, Germany has extensively imported capital (Deutsche Bundesbank, 1999).

Economically, there is nothing wrong about the reversal in the balance of payment situation. The estimated investment requirements in East

Germany of approximately 1 000 billion DM to catch up with the 1989 capital intensity of West Germany must be financed somehow. Part of the capital stock of around 500 billion DM that West Germany had accumulated abroad by 1990 could certainly be used for the East German adjustment process (Sinn and Sinn, 1994).

Table 2 gives an overview of public transfers from West to East Germany from 1991 to 1998. The bulk of these transfers obviously has come from federal government resources, either directly or as a share of federal labour office payments or statutory pension insurance schemes. Compared to these amounts, transfers from the 'German Unity' fund, the EU budget, and West German Länder and local authorities have been relatively low. Interestingly, gross transfers have continuously increased in the years after reunification, attaining a maximum level in 1996 where they have remained since. Table 2 also reveals that the gross transfers to East Germany to a large extent consist of cash transfers, and the greatest share has been used to establish a social safety net during the transition. Overall, only about one-third of the gross transfers to East Germany have been covered by revenues in this region, amounting to net transfers of around 140 billion DM in the years since 1995. It is this continuing high net public transfer level that indicates that the adjustment process in East Germany is far from completion.

The agricultural sector in East Germany also received high public transfers from West Germany. From 1991 to 1995 transfers from the federal budget amounted to around 11 billion DM. In the first years after reunification a major share of these transfers simply could be characterised as liquidity and income support whereas in the following years, transfers were used to extend the agricultural support system established in West Germany to the East (Boss and Rosenschon, 1996). The EU also contributed to the transfers to East Germany, although modestly. Table 2 shows that transfers have amounted to about 7 billion DM per year in recent years. Much of these transfers are not directly related to the reunification process but represent, for instance costs of extending the EU's Common Agricultural Policy (CAP) to East Germany. Hence, the agriculture sector may have particularly benefited from these European transfers. The EU has also launched special support programmes to support the adjustment process in East Germany. This region has been granted the status of an 'objective 1' region and from 1994 to 1999 about

Table 2:
Public transfers to East Germany (billion DM)

	1991	1992	1993	1994	1995	1996	1997	1998a	1991–1998
I. Gross transfers									
Total	139	151	167	169	185	187	183	189	1 370
Of which									
Federal Government	75	88	114	114	135	138	131	139	934
West German Länder and local authorities	5	5	10	14	10	11	11	11	77
'German Unity' Fund	31	24	15	5	–	–	–	–	75
EU budget	4	5	5	6	7	7	7	7	48
Federal Labour Office	24	24	14	18	16	12	16	14	138
Statutory pension insurance scheme	–	5	9	12	17	19	18	18	98
Of which									
Social security benefits	56	68	77	74	79	84	81	84	603
Subsidies	8	10	11	17	18	15	14	16	109

Table 2 continued

	1991	1992	1993	1994	1995	1996	1997	1998[a]	1991–1998
Investments	22	23	26	26	34	33	32	33	229
Cash transfers, not classifiable	53	50	53	52	54	55	56	56	429
II. Federal Government revenues in East Germany									
Total	−33	−37	−39	−43	−45	−47	−47	−48	−339
Of which									
Tax revenue	−31	−35	−37	−41	−43	−45	−45	−46	−323
Administrative revenue	−2	−2	−2	−2	−2	−2	−2	−2	−16
III. Net transfers									
Total	106	114	128	126	140	140	136	141	1 031
Memo item: Deficit of the Treuhand agency[b]	9	14	24	24	–	–	–	–	–

[a] Based on the Federal Government budget for 1998.
[b] Deficit adjusted for payments and reimbursements of interest by the Treuhand agency.
Source: Deutsche Bundesbank (1998)

27 billion DM of EU structural funds were transferred to East Germany. Rural development in East Germany has benefited from a transfer from Brussels of around 6 billion DM during this period, triggering additional private investments of around 11 billion DM (BMELF, 1998).

Figure 7:
Financial transfers between the Bundesländer, 1998

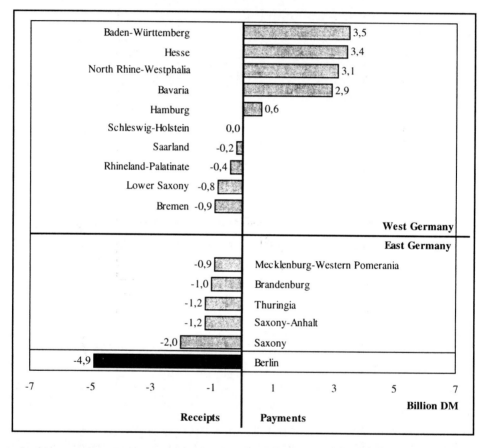

Source: BMF (1999)

The financial implications of the German reunification have resulted in a greater awareness of private and public financial restrictions in the economy. There is a serious and ongoing debate on cutting down budget deficits and public debt at all administrative levels. There is also rising concern about the extent and duration of financial transfers within the country. Baden-Württemberg, Bavaria, and Hesse have successfully questioned the fiscal transfer system between German Länder called 'Länderfinanzausgleich' which is a constituent of the federal structure of Germany. Figure 7 shows the net transfer flows between the Länder, and the figure reveals that today not only the West is subsidising the East, but also that the North is financed by the South.

The current debate on financial transfers with respect to the agricultural sector is somewhat similar to the general debate. The East German agricultural sector has received considerable transfers due not only to reunification, but also due to the reform of the CAP in 1992. Maintaining this transfer level has become a major concern for East German policy makers in the ongoing debate on further reform of the CAP. On the other hand, the agricultural support level is criticised more and more by the general public in Germany, reflecting the overall debate on the constraints and outlook for Germany's financial situation.

Concluding Remarks

Ten years after reunification the adjustment process of the East German economy is far from being completed. Today the GNP per capita in East Germany is only 56 percent of that in West Germany, and unemployment rates and public transfers continue to be high. There is an ongoing debate on the proper policy framework of the adjustment process and the role of markets in the transition process. Assessments of the current situation and future developments vary from a rather pessimistic to an optimistic view.

The agricultural sector has played a significant role in the adjustment process, and the process itself has affected the sector in specific ways. Despite enormous structural changes, huge dismissals of labour, severe financial constraints, and an weak overall economic environment, economic transition not only causes problems, but also offers opportunities. Today the most efficient farms in Germany can be found in the East, and average labour income is higher in the East than in the West. Despite the need for further adjustment and expectation of further liberalisation of the CAP, this is a promising vista.

References

Blanchard, O. (1996). Assessment of the Economic Transition in Central and Eastern Europe : Theoretical Aspects of Transition. *AEA Papers and Proceedings*: 177-122.

BMELF (Bundesministerium für Ernährung, Landwirtschaft und Forsten) (various issues). *Agrarbericht der Bundesregierung*. Bonn: Bundesregierung.

BMF (Bundesministerium der Finanzen) (1999). Bund-Länder Finanzbeziehungen auf der Grundlage der geltenden Finanzverfassungsordnung. *Das Bundesministerium der Finanzen informiert* No. 5/99. Bonn: Bundesregierung

Boss, A. and Rosenschon, A. (1996). Öffentliche Transferleistungen zur Finanzierung der deutschen Einheit : Eine Bestandsaufnahme. *Kieler Diskussionsbeiträge* 269. Kiel: Institut für Weltwirtschaft.

Burda, M.C. (1993). The Determinants of East-West German Migration : Some First Results. *Discussion paper series* 764. London: Centre for Economic Policy Research.

Burda, M.C. and Funke, M. (1995). Eastern Germany : Can't We Be More Optimistic? *ifo studien - Zeitschrift für empirische Wirtschaftsforschung* (41) No. 3: 327-354.

Deutsche Bundesbank (1998). Economic Conditions in Eastern Germany. *Monthly Report* April 1998: 41-54.

Deutsche Bundesbank (1999). Zahlungsbilanzstatistik August 1999. *Statistisches Beiheft zum Monatsbericht* 3: 6-7.

DIW (Deutsches Institut für Wirtschaftsforschung Berlin), IfW (Institut für Weltwirtschaft an der Universität Kiel), and IWH (Institut für Wirtschaftsforschung Halle) (1999). Gesamtwirtschaftliche und unternehmerische Anpassungsfortschritte in Ostdeutschland. In: http://diw-berlin.de/diwwbd/99-23-1.html (18.09.1999).

Häger, G. (1999). *Preise, Technologien und ökonomisches Verhalten als wesentliche Bestimmungsfaktoren in der Transformation*. Kiel: Vauk.

Häger, G., Kirschke, D., and Noleppa, S. (1999). *Ausmaß und Bedeutung von Marktkräften im Transformationsprozeß - Eine empirische Analyse zur Anpassung der Agrarproduktion in den neuen Bundesländern -*. Berlin: Humboldt-Universität zu Berlin.

Häger, A. and Hagelschuer, P. (1996). Einige soziale Auswirkungen der Transformation im Agrarsektor der Neuen Bundesländer. *Working Paper* No. 21. Berlin: Humboldt-Universität zu Berlin, Wirtschafts- und Sozialwissenschaften an der Landwirtschaftlich-Gärtnerischen Fakultät.

IAB (Institut für Arbeitsmarkt- und Berufsforschung der Bundesanstalt für Arbeit) (1998). Arbeitsmarkt Ostdeutschland: Angebot an Arbeitskräften bleibt weiter hoch. *IAB kurzbericht* No. 10. Nürnberg: IAB.

ifo (1999). ifo-Wirtschaftskonjunktur 4/1999. *Monatsberichte des ifo Instituts für Wirtschaftsforschung* (51) No. 4. München: ifo-Institut.

Koester, U. and Brooks, K. (1998). Agricultural Transition in the New Federal States of Germany. In: von Witzke, H. and Tangermann, S. (eds.). *Economic Transition in Central and East Europe, and the Former Soviet Union : Implications for International Agricultural Trade.* Berlin: Humboldt-Universität zu Berlin: 232–265.

Kirschke, D., Odening, M., Doluschitz, R., Fock, T., Hagedorn, K., Rost, D., and von Witzke, H. (1998). *Weiterentwicklung der EU-Agrarpolitik - Aussichten für die neuen Bundesländer.* Kiel: Vauk.

Pohl, R. and Ragnitz, J. (1998). Ostdeutsche Wirtschaft: Kein Grund zu Resignation. *Wirtschaft im Wandel* No. 7: 3–18.

Siebert, H. (1992). Die reale Anpassung bei der Transformation einer Planwirtschaft. *Arbeitspapier* 500. Kiel: Institut für Weltwirtschaft.

Sinn, H.-W. (1999). EU Enlargement, Migration, and Lessons from German Unification. *Working Paper* 182. München: CESifo.

Sinn, G. and Sinn, H.-W. (1994). *Jumpstart : the Economic Unification of Germany.* Cambridge: MIT Press.

StBA (Statistisches Bundesamt) (various issues). *Statistisches Jahrbuch für die Bundesrepublik Deutschland.* Stuttgart: Metzler-Poeschel.

Tietmeyer, H. (1999). *50 Jahre Deutsche Mark in Berlin - 50 Jahre Landeszentralbank in Berlin und Brandenburg.* Berlin: Landeszentralbank Berlin-Brandenburg.

Chapter 6

Rural Development in Germany – Issues and Policies

Eckhart Neander, formerly Institute of Structural Research, Federal Agricultural Research Centre, and Helmut Schrader, Institute of Farm Economics and Rural Studies, Federal Agricultural Research Centre, Braunschweig

Rural Regions in Germany

Definition of Rural Regions

Regions may be classified according to function or homogeneity. In the latter case characteristics of settlement structure as well as those of the structure of the economy may be used. 'Rural' regions are frequently characterized by a below average population density and the absence of larger conurbations, or by an above average proportion of the primary sector in land use, employment, and income formation.

In the Federal Republic of Germany (FRG), a method of clustering regions by settlement structure attributes came into use at the beginning of the 1980s, with population density and the existence of central places as characteristics for categorizing the 'Kreise' (administrative units roughly corresponding to counties in the U.K. and the U.S.A.) as types of regions. Kreise with a population density below 150 inhabitants per km² are called 'rural'. A further subdivision distinguishes between rural Kreise near large agglomerations (central places with 300 thousand or more people), those in areas with urbanizing tendencies (central places with between 100 and

300 thousand people), and those with no higher order central places in the vicinity.

Importance and Recent Development of Rural Regions

In 1996 rural regions, as defined above, embraced 60% of the area and 27% of the population of Germany with an average population density of 104 inhabitants per km^2, compared to 230 in all of Germany, and 418 in the non-rural regions. In the 'new' Länder (territory of the former German Democratic Republic), the proportion of rural regions was substantially higher with 79% of the area, 43% of the population, and an average population density of only 89 inhabitants per km^2.

Rural regions in West and in East Germany have experienced quite different types of development of population, employment, and economic strength during past decades (Schrader 1999; Becker 1997). Between 1980 and 1990 population numbers in rural regions of the original FRG grew at above average speed, solely due to net immigration from agglomerated and urbanizing regions, whereas natural population development was negative in all types of regions. In contrast, in the rural regions of the German Democratic Republic (GDR) population numbers declined during the same period, exclusively due to net emigrations (partly to larger cities, partly legally or illegally to West Germany), whereas the natural population development was positive. After the unification of West and East Germany in October 1990, the tendencies of population development in rural regions of West Germany continued. However in the new Länder, where the birth rate had previously exceeded the death rate, the trend reversed. Additionally net migration from rural areas accelerated. From the mid-1990s however birth rates have again started to climb, and for the first time rural regions near agglomerations or other cities show positive net migration figures. Thus, at least in the more densely populated rural regions in the southern and western parts of the new Länder, tendencies of population development have begun to approach those in rural regions of the original territory of the FRG.

In rural regions of the FRG the number of employed persons in the second half of the 1980s grew faster than average; unemployment rates were lower than average and decreased more quickly. For the territory of the former GDR comparable data on employment are not available for this time period, but the percentage of the population that was gainfully employed at the end of the 1980s was considerably higher than in the former FRG (75% against 45% of all persons in employable age). Rural regions even exceeded the average mainly due to a higher employment frequency of women. From 1990 to 1996 employment numbers in the old Länder continued to rise accompanied by falling growth rates in rural regions in contrast to agglomerated and urbanizing regions where employment figures had started to shrink and unemployment rates to rise. At the same time, in the new Länder employment figures fell dramatically by one-third, even faster in rural regions than in urban areas, and unemployment rates rose to 20% and more, women being particularly affected. However since the mid-1990s the employment situation has begun to stabilize.

In rural regions, average value added per employed person lags behind that of agglomerated and urbanizing regions by about 10 to 15% in both parts of Germany, although on a lower absolute level (about 60%) in the new Länder. Average tax revenues per inhabitant in rural regions fall short, even more than those of urban regions. For the second half of the 1990s data are not yet available. It seems, however, that disparities in productivity, available income, and tax power between rural and urban regions, which are at least partly due to differences in the structures of the respective economies, melt away only slowly.

Position of the Primary Sector

According to public opinion, 'rurality' of regions is frequently associated with predominance of the primary sector (agriculture, forestry, fisheries) in the regional economy. But this holds true in Germany only with respect to land use: in rural regions 53% of total land is used by agriculture, 30% by forestry. Only in the large conurbations are farm and forestry lands found in significantly smaller proportions. The contributions of the primary sector to employment and income formation in rural regions are much less impressive. At the end of the 1980s, in rural regions of the former FRG only about 7% of all gainfully employed people worked in agriculture, forestry, or fisheries with maximum figures of up to 15% in single Kreise, as compared to 3% for the whole territory. In contrast, in the former GDR the respective proportion was 10% for the whole territory, and between 15 and 30% in rural regions. This difference was caused not only by significantly lower labor productivity on farms in the GDR, but also by the fact that the large collective and state farms were, among other things, responsible for maintaining the technical and social infrastructure in rural regions which in Western countries is one of the tasks of the political community. After the unification of West and East Germany in 1990, farms in the new Länder had to reduce surplus manpower as soon as possible in order to survive under the completely changed legal and economic circumstances. Together with the breakdown of large segments of industrial production due to a lack of competitiveness, the result was the above-mentioned dramatic fall of employment numbers in rural regions of the new Länder. Already in the mid-1990s the proportion of the primary sector in employment and income formation in rural regions of the new Länder approximated those in the old Länder; in 1994 agriculture, forestry, and fisheries contributed only about 3.5% to gross value added in rural regions of both parts of Germany compared to about 1% in the whole country.

Table 1:
Relative position and development of population, employment, and economic power in different types of regions in Germany

Types of regions	Population Per Km² 1996	Population % Change 1990-96	Persons employed % change 1990-96	Unemployment rate % 1997	Unemployment rate %-pts change 1993-97	Gross value added per emp. person 1994	Tax revenue per inhab. 1995
						total Germany = 100	
Total 'old' Länder[a]	259	+4.6	-0.2	10.4	+2.3	109	112
Urban regions	415	+4.0	-1.3	10.8	+2.5	113	120
Rural regions, by type[c] 1	121	+8.5	+7.1	8.8	+1.9	93	90
2	116	+7.2	+5.6	9.3	+1.9	93	85
3	111	+5.1	+3.7	8.9	+1.8	92	85
Total 'new' Länder[b]	162	-3.3	-33.3	18.4	+2.9	67	56
Urban regions	435	-3.7	-29.2	17.9	+3.4	71	66
Rural regions, by type[c] 1	92	+1.4	-30.6	17.3	+2.4	58	50
2	99	-2.9	-41.6	19.9	+2.6	60	40
3	81	-4.7	-40.7	19.7	+2.0	60	41

[a] territory of former FRG; [b] territory of former GDR; [c] type 1 = rural Kreise near large agglomerations, type 2 = rural Kreise in areas with urbanizing tendencies, type 3 = rural Kreise with no higher order central places in the vicinity.
Source: Aktuelle Daten zur Entwicklung der Städte, Kreise und Gemeinden, Ausgabe 1998; Berichte des Bundesamtes für Bauwesen und Raumordnung, Band 1.Bonn 1998.

Public Policies for Developing Rural Regions in Germany

The following summary of public activities towards rural development in Germany since the end of World War II will be limited to those fields of policy which directly aim at influencing the regional and sectoral structures of the economy. Doubtless there are other fields of policy which more or less indirectly affect the relative economic position and development of rural regions (e.g. fiscal policy, labor market policy, education and research policy, transportation policy, military policy, farm market and price policy) but they will not be taken into consideration here (cf. Henrichsmeyer et. al. 1994).

West German Policies up to 1990

Up to the end of the 1960s The demographic and economic development of rural regions in the former territory of the FRG has not always been as positive as recently. After World War II, in many rural regions of West Germany, neither agriculture and forestry nor industry offered sufficient opportunities for gainful employment for a native population enlarged by great numbers of war refugees and expellees. Consequently incomes and living conditions lagged behind the rest of the economy, and people of working age left rural regions to find employment elsewhere. This was particularly true in rural regions without traditional industries, unfavorable natural or/and structural conditions for farming, as well as those along the 'iron curtain'.

Attempts to counteract the above mentioned developments concentrated on the creation of new and stable jobs outside agriculture and on the improvement of structural conditions within agriculture to stabilize family farming. According to the constitution of the FRG of 1949, the Länder were responsible for structural policies of any kind in their respective territories. But great differences in size, economic potential, and structural conditions existed between the Länder, as well as in the degree to which they were affected by the aftermath of World War II (destruction, inflow of refugees and expellees, iron curtain) which challenged the constitutional imperative to guarantee equivalent living conditions in the whole territory. Therefore the federal government started to intervene with programs to help the recovery, stabilization, and

development of disadvantaged regions, most of which were rural. These programs included (Breloh and Struff 1969):
- Beginning in 1951, investments in industrial enterprises and communal infrastructures in order to create new nonagricultural employment in so-called 'Notstandsgebieten' (depressed regions mainly in Schleswig-Holstein, Niedersachsen, and Bayern), in 1953 enlarged by the 'Zonenrandgebiete' (regions along the 'iron curtain'). Financed by the Federal Ministry of Economic Affairs.
- From 1951, programs to improve local natural and infrastructural conditions of agriculture (e.g. regulating the local water regime, land consolidation, road construction) and working conditions on farms (e.g. modernizing farm buildings and machinery) by offering financial support to individual farmers or responsible public bodies, in regions with particular structural disadvantages (e.g. regions bordering the North Sea coast or the Alps). Financed by the Federal Ministry of Agriculture.

In the course of time coordination of the respective activities between federal and Länder governments developed to a certain degree, but almost no formal coordination of programs took place at either level.

From the early 1970s to the late 1980s In 1969, as part of a comprehensive financial reform in the FRG, the partly complementary and partly competing activities of the federal and the Länder governments in the field of structural policies were given a new constitutional basis by creating so-called 'Gemeinschaftsaufgaben' (joint tasks). These tasks were directed towards the improvement of the regional structure of the economy (GRW) and for the improvement of the structure of agriculture and coastal protection (GAK), both of which were important for the development of rural regions. Committees consisting of representatives of the federal and the Länder ministries of economic affairs (with respect to GRW) or of agriculture (with respect to GAK) were organized, in which the federal representative was given an equal number of votes as those of the Länder combined. Each year the principles and design of programs and the distribution of financial means between programs and between Länder ('framework plans') are decided. Funding is supplied partly by the federal budget (50% in GRW, 60% in GAK other than for coastal protection, 70% for coastal protection) and partly by the Länder budgets. The Länder are responsible for the execution of the programs in their respective territories. Both joint tasks have experienced various changes since their implementation in the early 1970s.

The GRW joint task has undergone relatively few fundamental changes with respect to the core of activities promoted (Karl et. al. 1997). Investments are aimed at establishing, expanding, readjusting, or rationalizing industrial enterprises producing commodities or services tradable across the borders of the respective regions (including tourism) as well as extending or modernizing industry-related communal infrastructure. Investments are restricted to regions in which economic power is significantly below average or falls behind due to structural changes, with the objective of securing or creating jobs in these regions. Changes have been made, however, with respect to the regions included. At first these were predominantly rural regions, but the oil price explosion and structural crises in certain industrial branches (e.g. textiles, coal mining, steel processing, shipbuilding) along with rising unemployment beginning in the mid-1970s, led to corresponding adjustments of the criteria used for selecting the program regions. As a result there was a gradual shift from rural to urban regions with a high concentration of the most affected industries, and within the program regions financial means were directed more toward urban centers.

The promotion programs within the GAK joint task have always been offered countrywide, not restricted to certain (e.g. rural) regions except for special support for agriculture in less favored areas. But the contents of the programs and their relative weights have changed over time (Gießübel et. al. 1986; Henrichsmeyer et. al. 1994; Urff, von, et. al. 1996; Bundesministerium für Ernährung, Landwirtschaft und Forsten (ed.) 1998). In addition to the above mentioned activities aimed at improvement of local natural and infrastructural farming conditions, the GAK programs include the promotion of a wide range of activities including:
- on-farm investments to modernize and expand the income capacity of individual farms
- since 1974 direct 'compensatory' payments to farmers for continuing to farm in less favored areas
- improvements of marketing conditions for farm commodities through establishing producer organizations and investment aids to modernize marketing and processing enterprises

- establishment of central drinking water supply and sewage disposal installations in rural communities
- since 1977 renewal of private and public buildings and infrastructure in villages to increase their attractiveness ('Dorferneuerung')
- protection of the coasts along the North Sea and the Baltic Sea from tidal damage.

This list mentions only the more important activities with respect to financial input. The participation in the different programs in terms of this input has varied over time as well as between the Länder.

Up to the mid-1960s, structural policies for agriculture in the FRG were aimed mainly at maintaining as many family farms as possible by attempting to improve their natural and infrastructural production conditions. The fast growth of the general economy since the mid-1950s with a steadily, rising demand for workers in industry, trade, and services together with the significant income disparity between families operating small, full-time farms and most of the working population outside agriculture encouraged more and more farmers to take up non-farm jobs, giving up their farms or turning them into part-time businesses. Correspondingly the size and production capacity of the remaining farms increased, intensified by their adoption of technical changes. In the second half of the 1960s the realization of the inevitability of this structural change in agriculture caused the federal government to restrict on-farm investment aid to full-time farms capable of attaining an income equivalent to the average non-farm income within four years, and to offer other farm families programs for facilitating the transfer to non-farm jobs or the taking of early retirement. This new orientation of on-farm investment aid, introduced into the GAK first framework plan in 1973, was, however, gradually watered down in the following years.

A further change in the structure of the GAK programs took place in the mid-1980s when, as a political compensation for cuts in the European Community (EC) farm price support, the number of so-called less favored areas entitled to receive direct compensatory payments was sharply increased to more than 50% of the agricultural area in the FRG. This inflated the share of these payments in total GAK expenses from less than 10% to more than 25%. Between 1972 and 1989 the federal and Länder budgets together annually contributed about 690 million DM for the GRW and about 2.2 billion DM for the GAK in the original territory of the FRG. The proportion of financial means allocated to rural regions is unknown. The figures clearly demonstrate the predominant financial means addressed to the primary sector by structural policy. As in the 1950s and 1960s, there was no effective coordination between the two joint tasks.

The period between the early 1970s and the late 1980s was also characterized by a steady expansion of activities of the EC in structural policies. On one hand, the EC participated in co-financing national programs to promote structural adjustments, but also developed Community-wide programs with this objective. Thus, from 1972 on, national programs of farm modernization, retraining of the farm population, early retirement of farmers, strengthening market structures for farm products, and maintaining farming in less favored areas became eligible for co-financing by the guidance section of the European Agricultural Guidance and Guarantee Fund (EAGGF), if they were in accordance with respective EC directives and regulations. The FRG federal government made ample use of this funding. In the following years the number of promotion activities eligible for co-financing was expanded. In 1975 the European Regional Development Fund (ERDF) was established and, after first being restricted to co-financing national regional development projects, in course of time it was also enabled to evolve Community-wide programs. On the other hand the European Commission strengthened its control of national programs with respect to their effects on competition on the common market, thus affecting the expansion of the GRW joint task during the 1980s.

From 1988 to 1990   In the Single European Act of 1986, which represented an important further step towards deepening the cooperation within the EC, the European Community's obligation for internal economic and social cohesion was established, thus strengthening its responsibility for regional structural policies. This led to a comprehensive reform of the Community's structural funds in 1988, mainly including:

- a concentration of the three structural funds (the EAGGF-guidance section, the ERDF, and the European Social Fund) on six objectives: 1, 2, 3, 4, 5a, and 5b
- a coordinated use of the funds, partly together with the European Investment Bank

- a safeguard that structural funds would truly be used as an addition to and not a substitute for funds provided by public and/or private sources in the Member States (principle of 'additionality')
- participation of the European Commission, Member States, and regional authorities in planning, executing, monitoring, and evaluating the structural measures co-financed by the structural funds (principle of 'partnership').

In 1988 only the objectives 5a and 5b became relevant for rural regions in the FRG. Objective 5a aimed at speeding up the necessary adjustments of agricultural structures to comply with the reforms of the Common Agricultural Policy by means of the guidance section of the EAGGF horizontally, i.e. countrywide. Objective 5b focused on facilitating the development and structural adjustment of rural areas by means of all three structural funds, and was restricted to regions with an above average proportion of employment in agriculture, a below average gross value added per working unit in this sector, and a low gross domestic product per capita related to the national average. Objective 5a covered the traditional instruments for EC structural policy in agriculture and was executed in the FRG through co-financing 25% of the promotion of on-farm investments, of the maintenance of farming in less favored areas by direct payments, of creating producer organizations, and of investments to modernize agricultural market structures within the GAK joint task. Objective 5b provided a much broader variety of approaches. Besides the modernization and diversification of enterprises in agriculture and forestry and their respective processing industries, it covered the development and modernization of enterprises in branches outside the primary sector including tourism. Furthermore it included the improvement of rural infrastructure, the natural environment and the landscape, and the training and qualification of people working in these branches or elsewhere. From 1989 to 1993 objective 5b program regions in the FRG covered 21% of its territory and 7% of the total population. These regions were concentrated in the Länder Bayern and Niedersachsen and received about 1.1 billion DM from the structural funds.

East German Policies up to 1990

The foremost objective of rural development in the GDR was the approximation of working and living conditions in agriculture and in rural regions to those in industry and urban centers in accordance with the ideology of the supremacy of the industrial working class. The realization of this objective was, however, severely hampered, particularly in the northern GDR which was traditionally far less developed with respect to population density, availability of central places, infrastructural endowments, and employment opportunities outside agriculture than the southern part. This handicap was sharpened after World War II by separation from important metropolitan functions of the cities of Hamburg in the West and Stettin in the East and by an above average load of refugees and expellees from former German territories and East European countries. In view of these problems, the GDR authorities pursued their objectives in various ways and with varying intensity over time (Grimm 1992; Zierold 1997).

Extremely large agricultural enterprises were created in the course of collectivization in the 1950s and 1960s and with the introduction of industrial production and organization methods in the 1970s. These enterprises frequently covered the territory of several villages and not only established new buildings for livestock herds and machinery, but also for the farm workers and their families, leaving the old buildings for private use or to decay. The enterprises were also responsible for supporting the rural communities in maintaining the technical and social infrastructure to supply the territory's population with everyday goods and services. These were generally distributed from the enterprise management site.

In selected urban areas or new settlements in rural regions, in the 1950s and 1960s attempts were made to improve employment and housing opportunities by establishing industrial branches and constructing new housing. Public transit services enabled working people to commute from their homes in small, rural villages to urban areas. In this way, the authorities partly succeeded in moderating the locational disadvantages of sparsely populated and poorly structured rural regions. Nevertheless they were unable to completely prevent the out-migration from small villages and towns to larger conurbations driven mainly by better housing and employment opportunities, but also by the legal and illegal migration to West Germany.

Policies in United Germany Since 1990

The economic, monetary, and social union reached by the two German states in July 1990, and their final unification in October 1990, resulted in far-reaching changes for rural regions in the new Länder. To increase labor productivity in accordance with the new market and policy conditions, old but transformed as well as newly established farm enterprises had to get along with as few workers as possible and release the rest. A high proportion of the industrial enterprises established in rural regions under the GDR authorities were forced to close down under the new conditions due to a lack of competitiveness. Thus, employment shrank even faster in rural regions than elsewhere. In order to prevent a mass out-migration of labor from rural regions, interventions were needed to save and stabilize the remaining jobs and to create new ones by modernizing their economically relevant structures.

With the unification, the joint tasks GRW and GAK came into effect in the new Länder. Their complete territory became eligible for the GRW programs with promotion conditions significantly more attractive than in the old Länder while the GRW promotion in the old Länder was reduced, particularly the special one along the former German internal border. An additional financial volume of 8 billion DM annually was dedicated to the GRW in the new Länder. Until 1993, investment aid for establishing or modernizing industrial enterprises (including tourism) predominantly went to urban agglomerations, whereas rural regions participated more in the promotion of investments to modernize infrastructure. Up to 1993 in the GAK joint task, an additional 1.6 billion DM were provided annually by the federal and the Länder budgets for the new Länder. The funds were particularly used for the support of farm establishment and modernization, the maintenance of farming in less favored areas, and the inducement of investments to modernize agricultural marketing structures, drinking water supply and sewage disposal installations, and the renewal of villages. The contributions of the old Länder in favor of the new ones were supported by the provision of additional financial means from the EC structural funds: 2 billion DM annually from 1991 and 1993,

50% from the ERDF, 30% from the ESF, and 20% from the EAGGF-guidance section, based on a joint promotion concept.

There were additional efforts to support and accelerate the necessary adaptation of the economic and social structures in the new Länder to the new conditions, especially in rural regions (e.g. liquidity aid to farms to help them cope with the new and somewhat destructive market conditions for their products, a special program to modernize and supplement the road and rail network to improve accessibility of the new Länder). All these activities were accompanied by intensive attempts to ease labor market adjustment through programs providing retraining for persons affected or threatened by unemployment in primary labor markets, creating secondary labor markets for such persons, and offering early retirement.

From 1994 to 1999, the second phase of the application of the reformed EC structural funds, several changes were introduced to the policies for developing rural regions (Neander et. al. 1997). The new Länder, with a per capita domestic product of less than 75% of the EC average, became fully eligible for EC structural funds according to objective 1 (figure 1, see p. 133). Among the six main fields of support for objective 1 regions, program 6 is dedicated to agriculture, rural development, and fisheries and includes five subprograms. Subprogram I covers aid for investments on farms as well as for improvements in marketing and processing farm products. Subprogram II supports activities for rural infrastructure, investments in non-farm enterprises and rural tourism, land consolidation, road construction, and village renewal. Subprogram III is directed towards improvement of the environmental conditions of farming and forestry, the promotion of small crafts and industries, and support of direct marketing of farm products. Subprogram IV offers technical aid for the preparation, execution, monitoring, and evaluation of the above mentioned subprograms. Finally, a fifth subprogram provides promotion activities for adjusting and modernizing the fishery sector. Up to 75% of total expenditures and at least 50% of public means spent for these subprograms may be financed by EC structural funds. The funding for objective 1 for the period 1994 to 1999 is expected to total about 26 billion DM, of which 6.2 billion DM are assigned to program 6. Each of the new Länder had to submit an operational program for the use of these financial means containing a survey of the specific measures to be applied and a forecast of their expected effects on employment and incomes (Grajewski et. al. 1994).

In the old Länder the promotion activities under objective 5a have been extended to the adjustment and modernization of fisheries using a newly created financial fund, FIAF. The areas eligible for promotion of rural development under objective 5b have been expanded by 75% compared to the first phase (figure 1), and the funds earmarked for this objective have been increased to 2.3 billion DM: 42% from the EAGGF-guidance section, 39% from the ERDF, and 19% from the ESF. For objective 5b promotion, the old Länder also had to submit operational programs.

In addition to the programs already mentioned, so-called 'Community Initiatives' may also be supported by EC structural funds. For rural regions the relevant Community Initiatives are LEADER and INTERREG; the first providing support to local projects incorporating exemplary and innovative strategies for rural development, the second providing support to cross-border cooperative activities. For 1994 to 1999, a total of 395 million DM is planned for LEADER II projects (175 million DM in the new Länder, 220 million DM in the old Länder) and another 920 million DM for projects under INTERREG II in both parts of Germany.

Policy Changes After 1999

In March 1999, the European Council of heads of governments decided upon a further reform of EC structural policies for the time period 2000 to 2006. The new reform has become inevitable in view of the need for changes in common agricultural and structural policies and to incorporate aid for structural adjustments in acceding Central and Eastern European countries which would draw from the Community's budget without significantly contributing to it. The most important changes relevant for rural development policies are as follows:

- The objectives of the EC structural funds are reduced to three. Objective 1 still supports the development and adjustment of regions whose economies lag significantly behind (per capita gross domestic product less than 75% of the EC average). Objective 2 is now directed towards economic and social readjustments in regions with structural problems, which may include industrial and urban regions as well as rural regions and those with an important fishery sector. Objective 3 contains the adjustment and modernization of education, training, and employment systems and policies outside objective 1 regions. All other objectives, including objectives 5a and 5b, relevant for the development of rural regions have been cancelled or partially incorporated in the new approach.

- The promotion of rural development now includes not only measures hitherto parts of objectives 5a and 5b programs, but also measures introduced during the late 1980s and early 1990s as part of the reforms of the common agricultural market policy. These policies aim to reduce supply pressure on EC farm commodity markets, accelerate structural adjustments, improve environmental effects of agricultural production methods (namely promoting reforestation of farmland), offer early retirement to farmers and farm workers, and support environmental programs. All these measures will now be co-financed by the EAGGF-guarantee section.
- With the exception just mentioned, the promotion of rural development in objective 1 regions will continue to be financed jointly by the three structural funds (ERDF, ESF, and EAGGF-guidance section) according to EC regulation no. 1260/1999 of the structural funds. Outside objective 1 regions, however, the promotion of rural development (including most of the measures previously supported under objectives 5a and 5b as well as those mentioned in the preceding paragraph) may be applied countrywide without regional restriction according to EC regulation no. 1257/1999, but have to be financed by the EAGGF-guarantee section. Thus programs compete against EC market policies for limited resources according to the 'agricultural guideline' of the EC budget.
- Regions hitherto supported under objective 1 or objective 5b but eliminated from these categories after 1999 will receive transitional funding for six years from the EC structural funds, for which special financial resources have been reserved.

In Germany, the new Länder will remain objective 1 regions after 1999 with the exception of the eastern part of Berlin. Thus rural regions in the new Länder will remain under an EC promotion regime roughly similar to the 1994 program. In the old Länder, only a very few rural regions will qualify for promotion under objective 2, since these regions will have to prove not only a population density of less than 100 per km² or a proportion of working people in agriculture of at least twice EC average, but also a three year unemployment rate above the EC average or a loss of population since 1985. These conditions are not typical in rural regions in West Germany as previously discussed. Instead, the majority of rural regions will be eligible for funding according to EC regulation no. 1257/1999, and Länder governments will have to develop corresponding plans to enable them to qualify for co-financing by the EAGGF-guarantee section, and be permitted to support rural development with federal and Länder funds.

Evaluation Results of Rural Development Programs in Germany

Characterization of the Evaluation Approach

In the FRG, substantial, systematic evaluation of national policy measures to promote rural development hardly existed before the end of the 1980s. Partial investigations regarding the pattern and development of selected measures led to the following conclusions:
- based on the supply of financial means, measures provided to improve structural conditions in the primary sector were long given greater weight than those for promotion of employment in the secondary and tertiary sectors
- prevalent agrarian structural policy measures were predominantly focused on the improvement of natural and infrastructural local conditions of agriculture and forestry and to a minor degree on the promotion of necessary structural adjustments of appropriate farm operations
- aside from striving for an increase in employment and income formation by reallocating capital, labor, and know-how between sectors and regions, the above-mentioned structural policy measures almost always, although with varying intensity, tried to directly influence income distribution in favor of certain sectors,

social groups, or regions, thus frequently weakening their allocational efficiency
- efforts toward concentrating financial funds on regions with the highest requirement of structural adjustments were repeatedly softened during phases of plentiful financial means
- effective coordination of political structure measures in and outside the primary sector occurred at neither the Länder, federal, nor EC levels.

Only after the reform of the EC structural funds of 1988 did a systematic approach for the evaluation of supporting measures co-financed by these funds become mandatory for monitoring and impact assessment.

In the following section some results derived from evaluation studies, carried out for rural development programs in objective 5b regions in West Germany (since 1989) and objective 1 regions in East Germany (since 1991), are summarized (Schrader 1994; Schrader et. al. 1998). In accordance with existing EC regulations, evaluation of rural development programs in Germany cover the following main topics: assessment of the development strategy in coherence with the framework conditions of national and European structural policies, financial and physical realization of the support measures, administrative and organizational aspects of program implementation and assessment of impacts and efficiency, and synergy effects between measures of the rural development policy.

Characteristics of the Development Strategies

Usually each program consists of three subprograms or development components, each subprogram financed by a single EC structural fund. The financial resources are allocated between individual programs of the Länder, corresponding to their particular problem situation. The three subprograms pursue the following aims: 'diversification of the farm sector', co-financed by the EAGGF; 'development of nonagricultural sectors', co-financed by the ERDF; and 'development of human resources', co-financed by the ESF.

The main emphasis in German rural development programs is on the renewal and development of villages and consolidation of farmland, the promotion of employment and income formation outside agriculture, the qualification of persons for new employment opportunities by vocational training, and specific measures for environmental protection and landscape preservation. For example in German objective 5b programs, more than two-thirds of total public expenditures have been assigned to measures that take into account the regional disparities in labor market conditions and living conditions in the countryside as well as the increasing preference of society as a whole for improvements in environmental conditions.

Financial and Physical Target Achievement

In general, the implementation of financial interventions by the three structural funds involved in the support of rural regions in Germany, measured by respective commitments and payments, has, during the recent programming period of 1994 to 1999, proceeded in accordance with the targets set forth in the programming documents. According to financial records, only a few measures have lagged behind due to initial uncertainties about consistency with national regulations and financial procedures for their implementation. Major shifts between the funds have not been considered by the responsible authorities. In a few cases it has been recommended that endeavours be intensified for a better flow of information regarding support conditions to final beneficiaries and the granting procedures in the lagging regions.

The availability of physical indicators of program realization and effects varies between evaluation reports. Although a large number of indicators has been provided in evaluation studies, the comparability of specified data between the Länder programs is very limited. Many of the quantified indicators for physical program achievement can be interpreted only in a local context. In general, indicators are not suitable for aggregation with the intent of assessing the overall effectiveness of measures. As for quantitative data for target achievement presented in evaluation reports, the available information indicates a sufficient degree of target attainment. There remains a challenging obligation to define and select more uniform and consistent criteria for quantitative target achievement with the intent to improve the comparability of evaluation results.

Assessment of Program Management and Implementation Procedures

Summing up the recommendations from the different evaluation studies, the following main issues regarding program management and implementation procedures have been identified:
- Most frequently, improvement of the existing monitoring schemes and the actual use of these systems for program guidance and required adjustments to changing conditions is recommended.
- Participation of authorities and actors at the regional and/or local level could be strengthened in some areas. In a number of regions target-oriented attempts serve as an example for other 5b regions with regard to integration, local networks, motivation, and expert guidance to final beneficiaries provided by rural development groups and local agencies (e.g. by agencies for enhancement of regional economic activities as in the LEADER Community Initiative, particularly aimed at stimulating innovative measures).
- Generally the programming approach of the EC structural policy has been appreciated as an effective framework for administrative guidance of development strategy. Although for the chosen evaluation approach, more qualified knowledge of the sense and purpose of evaluations could improve the quality of impact assessment and the comparability of evaluation results.

Assessment of the Impact and Efficiency of Measures

During the course of program implementation, changes in the national economy have considerably affected the economic conditions in the supported regions. Thus the identification of the quantitative partial economic impacts of the measures to be evaluated has been hampered, but direct effects of the programs have been demonstrated in the evaluation studies that are available for a variety of measures, partially from results of measure-specific case studies. Employment conditions in the support regions have been improved to a degree by newly created jobs, but it has always been difficult to isolate these job-creating effects from developments which would have taken place without the programs. Generally, the expected long-run and side effects could not be confirmed with sufficient certainty by the available data due to the limited time the support programs have been in action.

Furthermore, the evaluation studies indicate that the value added by the EC structural fund interventions has been largely a result of an increase in the national funds for the regions concerned, whereas outside these regions promotion has stagnated or decreased. Thus a concentration of means in favor of support regions was achieved in accordance with the EC principle of 'additionality' by co-financing the EC structural funds combined with national means. The measures have contributed to new employment and income alternatives in and outside the primary sector and are regarded as meaningful support for facilitating the process of structural change in agriculture. In addition, support schemes for diversification, extensification, and natural protection measures within the framework of the programs may contribute somewhat to reducing surpluses on farm commodity markets resulting from past common agricultural market policies. Synergy effects between different activities of the programs have been identified in some cases; for example, by the promotion of rural tourism in connection with improvement of infrastructure, marketing activities, and vocational qualification. Those appropriate synergy effects cannot be generalized, however, and are only of small importance in relation to the volume of financial means. Within the framework of the programs a considerable proportion of environmentally relevant measures have been executed which contribute to the preservation of nature and improvement of landscape.

References

Becker, H. (1997). Dörfer heute – Ländliche Lebensverhältnisse im Wandel 1952, 1972 und 1993/95. *Schriftenreihe der Forschungsgesellschaft für Agrarpolitik und Agrarsoziologie e.V.* 307, Bonn.

Breloh, P. and Struff, R. (1969). Zur Frage der regionalen Förderprogramme in der Bundesrepublik Deutschland. *Berichte über Landwirtschaft* 47(2): 348-373.

Bundesministerium für Ernährung, Landwirtschaft und Forsten (ed.) (1998). 25 Jahre Rahmenplan der Gemeinschaftsaufgabe 'Verbesserung der Agrarstruktur und des Küstenschutzes'. Bericht über die Fachtagung vom 14. bis 16. Juli 1998 in Schwerin. Bonn.

Gießübel, R. and Spitzer, H. (1986). Federal Republic of Germany: Rural development under federal government. In M. Whitby (ed.), *Rural Development* in Europe: Some Surveys of Literature, Special Issue, *European Review of Agricultural Economics* 13(3): 283-307.

Grajewski, R., Schrader, H. and Tissen, G. (1994). Entwicklung und Förderung ländlicher Räume in den neuen Bundesländern. *Raumforschung und Raumordnung* 52(4/5): 270-278.

Grimm, F.-D. (1992). Ländlicher Raum und ländliche Siedlungen in der Siedlungs- und Raumordnungspolitik der ehemaligen DDR. In G. Henkel (ed.), *Der ländliche Raum in den neuen Bundesländern. Essener Geographische Arbeiten.* 54, Paderborn: 1-6.

Henrichsmeyer, W. and Witzke, H. P. (1994). Agrarpolitik, Band 2: Bewertung und Willensbildung. Stuttgart: Ulmer.

Karl, H. in collaboration with Krämer-Eis, A. (1997). Entwicklung der regionalen Wirtschaftspolitik in Deutschland. In H. H. Eberstein and H. Karl (eds.), *Handbuch der regionalen Wirtschaftsförderung.* Köln: O. Schmidt, A II 1-58.

Neander, E. and Schrader, H. (1997). Regionale Wirtschaftsförderung in ländlichen Räumen. In H. H. Eberstein and H. Karl (eds.), *Handbuch der regionalen Wirtschaftsförderung.* Köln: O. Schmidt, B VII 1-38.

Schrader, H. (1994). Impact assessment of the EU structural funds to support regional economic development in rural areas of Germany. *Journal of Rural Studies* (Special edition), 10(4): 357-365

Schrader, H. and Tissen, G. (1998). Förderung ländlicher Entwicklung durch die Europäischen Strukturfonds in Deutschland - Zwischenbewertung der Ziel-5b-Politik in Deutschland 1994 bis 1996. Endbericht für GD VI der Kommission der Europäischen Gemeinschaften. Braunschweig: http://europa.eu.int/comm/dg06/rur/eval/interim/de/text.pdf.

Schrader, H. (1999). Tendenzen und Perspektiven der Entwicklung ländlicher Räume. In P. Mehl (ed.), *Agrarstruktur und ländliche Räume - Rückblick und Ausblick.* Landbauforschung Völkenrode, Sonderheft 201, Braunschweig: 213-239.

Urff, W. von, Boisson, J. M. et. al. (1996). Regional Aspects of Common Agricultural Policy - New Roles for Rural Areas. Hannover: Akademie für Raumforschung und Landesplanung.

Zierold, K. (1997). Veränderungen von Lebenslagen in ländlichen Räumen der neuen Bundesländer. - In A. Becker (ed.), *Regionale Strukturen im Wandel.* Beiträge zu den Berichten der Kommission für die Erforschung des sozialen und politischen Wandels in den neuen Bundesländern e.V., Band 5.1, Opladen: 501-567.

Rural Development in Germany

Figure 1: German regions eligible for support by EC structural funds according to objectives 1 and 5b, 1994-1999

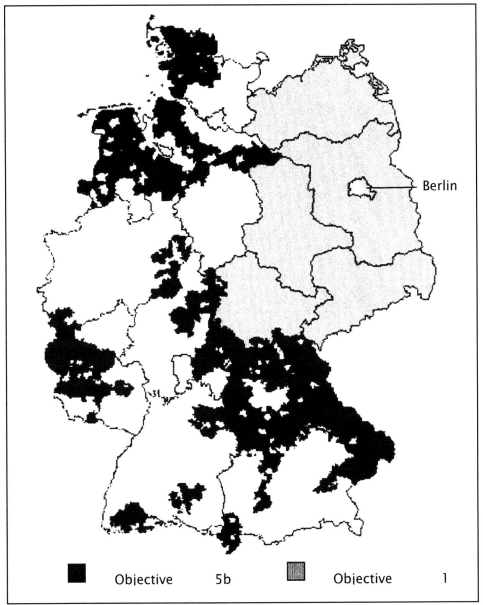

Source: Own presentation; delimitation is municipality-based according to specification by European Commission.

Chapter 7
Social Policies for German Agriculture

Konrad Hagedorn and Peter Mehl
Humboldt University of Berlin and Federal Agricultural
Research Centre Braunschweig

Introduction

After World War II the Federal Republic of Germany (FRG), France, Italy, and Luxembourg made specific social security provisions for their agricultural populations (Eggers, 1980). A simpler institutional solution would have been to integrate farmers into the general scheme, as did the United Kingdom, Denmark, and The Netherlands. In all fifteen member countries of the EU, however, agricultural employees and self-employed farmers are covered by comprehensive compulsory insurance schemes (Winkler, 1992; Mehl, 1997b).

Social security for the working population in the agricultural sector of the FRG is organised into two systems. Self-employed farmers and their families are insured in a special system which comprises an old age pension insurance, health insurance, long-term care insurance, and accident insurance. The system has evolved to meet the needs of social protection of self-employed farmers and their families (see Hagedorn, 1982a; 1986; Mehl, 1997a). With the exception of accident insurance, which is the responsibility of the employer, farm workers or employees on agricultural enterprises cannot participate in this special systems for farmers and their families. As their needs of social protection are different, they are insured under the statutory systems for the general working population.

Although it covers principally the same risks, in many respects the social security system for farmers does not function according to the precepts of general social policy. Because social security policy for agriculture in the FRG is largely subsidised, increasing strain on the federal budget requires reform. This is particularly true since German reunification, as the very heterogeneous structure of the agricultural sector that evolved in East Germany differs considerably from the West German family farm structure (see Hagedorn and Beckmann, 1995, 1997; Hagedorn, 1997; Beckmann 2000; Forstner and Isermeyer, 2000). Hence, income policy that subsidises the special social security system of self-employed farmers (family farms and partnerships) leads to unequal treatment compared with those farm enterprises (e.g. cooperatives, limited companies, joint stock companies) in which the labour force mainly consists of workers and employees.

For this reason, it is difficult to understand the present problems and peculiarities of the German social security system for the agricultural population without having some information on social issues and social policies related to reunification. Accordingly, we will first outline how the German government has dealt with social problems in the agricultural sector of former East Germany caused by the process of decollectivisation and privatisation. These measures were mainly part of the general transformation policies. In the following sections, we will deal with social security policies for agriculture, since, measured by expenditures, such policy 'has evolved to the most important pillar of agricultural policy on the national level' (Deutscher Bundestag, 1999, p. 2). Our focus is on one aspect, financing the social security system in agriculture (see also Hagedorn, 1987), although other aspects are also in need of reform (Hagedorn, 1982a; Mehl, 1997a). We will begin with an overview of the structure of the current system and its shortcomings, and then discuss the financial and distributive effects of the system. Reform is needed because federal subsidies are being used in a way that can no longer be justified politically. The results of the analysis show why these funds should be withdrawn from the farmers' social security scheme and instead used for adjustment policies to achieve structural and ecological changes, temporarily subsidising income for cushioning such changes.

Social Policies in the Transformation Process

Particularly in the East German agricultural sector, the very rapid transition from a planned economy to a market economy led to drastic job cuts (Fink et al., 1994, p. 282). Due to the new conditions, East German cooperative farms had to manage an enormous reduction in the labour force in a very short period of time, a difficult task (Figure 1). In 1989, about 850,000 people were working in East German agricultural enterprises; by the end of 1991, there were 300,000, about 50% of them part-time. In 1995, only 162,000 people were still working in former East German agricultural enterprises. Compared with the agricultural sector in other former Soviet bloc countries, this was an extremely rapid decline.

In contrast to other Central and Eastern European countries (CEECs) in transition, a bundle of government programmes have been adopted which contributed much to easing this process. Due to massive initial financing and delegation of many employees from West German institutions as counsellors, trainers, and instructors, a functional new labour administration was very quickly created in East Germany and was in effect at the time of political reunification (see Tegtmeier, 1993, p. 55-67). Figure 1 shows the status of persons at the end of 1991 who had lost their jobs in 1989:

- 120,000 (14.1%) had found another job
- 175,000 (20.6 %) aged 55 years and older had retired or taken early retirement, receiving 65% of their last net wage for a maximum of 5 years
- 105,000 (12.4%) were in additional vocational training, retraining, or job creation programmes

The early retirement schemes and the job creation measures provided considerable relief for the labour market. Vocational training and retraining helped the working population to adapt to changed demands. In addition, people in Germany are entitled to unemployment benefits or unemployment assistance; the latter can be claimed after unemployment benefits have expired. Payments are 60% to 67% (unemployment assistance 53% to 57%) of former earnings and are paid for up to 12 months, for older workers up to 32 months.

Obviously the financial dimensions of these support measures are enormous. Transfers from the federal budget to former East Germany amounted to 615 billion DM from 1991 to 1995, approximately 40% (215 billion DM) on social policies, mainly the labour market measures mentioned above. Additionally, contributors to the social insurance systems of the FRG were transferring another 140 billion DM to finance deficits in former East Germany. These considerable transfers resulted from the high unemployment rates, the introduction of the FRG benefit entitlement rules (Deutsche Bundesbank, 1996, p. 20f), special early retirement schemes, and minimum pensions.

Figure 1:
Decrease in Working Population in Agriculture in the New Federal States

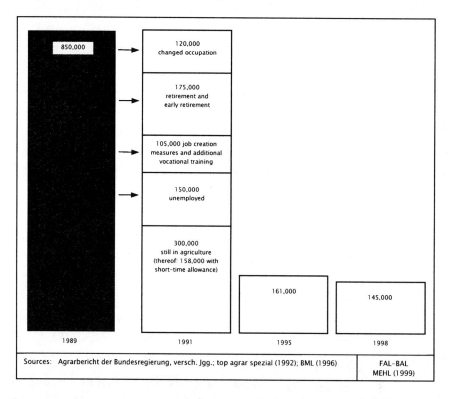

Due to these social programmes for cushioning the changes, former East Germany found itself in a unique situation which gave it, in comparison with other CEECs in transition, a rather privileged position. However, this general policy framework of transition could only be achieved because transformation in East Germany was almost without exception based on adopting West German institutions. As a consequence, employees of cooperative farms or other corporate enterprises were integrated into the FRG general statutory insurance systems for workers and employees; self-employed farmers and their families on new or reestablished family farms or partnerships in former East Germany came under the special insurance schemes for farmers.

The Social Security Systems for Farmers

The main feature of social security policy in the agricultural sector is that the system aimed not only at social security but also at structural and income policy goals. A farmer who reaches the age of retirement can only claim his pension benefit if he has transferred his farm to his successor.[1] The social security system in agriculture is largely financed by state subsidies in order to reduce contributions that otherwise had to be paid by farmers. Hence, agricultural social policy, still the responsibility of the national government, contains redistribution mechanisms that serve as a complement to the agricultural income policies of the European Union. Compared with the social security schemes for workers, the agricultural system has several additional particularities that have been designed to meet the special needs of social protection for self-employed farmers and their families (Table 1).

Origins and Development of the System

In the FRG, the social insurance system in agriculture has been expanded since 1957. At that time, policy makers decided to supplement the agricultural accident insurance which existed since 1886 with old age pensions for farmers, in an attempt to alleviate the social problems caused by accelerated structural change in agriculture. Low incomes were forcing many farmers who had not made adequate provisions for their old age to continue farming beyond retirement age, simply to secure a livelihood. The problem was compounded by the breakdown in the traditional form of security provided by the extended family. As a result, the transfer of farms

---

[1] Initially this farm transfer clause was a justifiable instrument of structural policy, because farmers were delaying transferring their farms. This justification has been weakened, as this is no longer a problem (see Hagedorn, 1981a).

from one generation to the next stagnated. A solution was urgently needed for both social and structural reasons.

The old age pension scheme for farmers ('Altershilfe für Landwirte') started with very low pensions. In the 1960s and 1970s, however, the level and scope of benefits were amended and increased considerably (Hagedorn, 1986a). In 1972, health insurance for farmers was introduced, providing almost the same benefits as the statutory health insurance for workers and employees. Long-term care insurance was introduced in the FRG in 1995. In this way, with the exception of unemployment insurance, social security for farmers covers the same risks as the general statutory social security systems.

Target Group

Insurance under the agricultural old age pension scheme is compulsory for all self-employed farmers whose farms fulfil the legal 'minimum size' requirements.[2] In accordance with the act to reform the social security system in agriculture (Gesetz zur Reform der agrarsozialen Sicherung – ASRG 1995), since 1994 spouses of farmers have been compulsorily integrated in the agricultural old age pension scheme, based on the argument that they play a vital role in running family farms and hence, are entitled to benefits in case of disability or old age. Farmers' successors and other family members employed on the farm together with their spouses could also participate in the scheme (Hagedorn, 1982a), but at 50% of the level of the contributions and benefits for farmers. Membership in agricultural health insurance is limited to farmers and their families; agricultural employees (i.e. managers and workers in former cooperatives or their legal successors, employees in family farms and partnership farms) remain in the general statutory systems. Agricultural accident insurance, like other branches of statutory occupational accident insurance in the FRG, is considered to cover employers' liability and therefore is funded by their contributions.

This includes family farms and partnerships as well as the other legal forms of agricultural enterprises. This means that the entire working

---

[2] The minimum size is regionally fixed by the agricultural old age pension funds; on average, about 4 ha.

population in the agricultural sector is covered by the agricultural accident insurance system.

Benefits

There are only minor differences between the health care and long-term care insurance benefits provided by the agricultural and the general statutory systems. Greater differences are seen in old age pensions and accident insurance coverage. This is because the benefits for wage earners and salaried employees from accident insurance and the old age pension scheme are to replace wages lost due to old age or in case of reduced earning capacity. The benefits depend on previous earnings, as protection from the financial consequences of the above-mentioned incidents and help them maintain their standard of living. In contrast, pensions and accident insurance benefits for farmers take into account that retired farmers customarily receive a part of their income from their successors. The main difference between the two pension formulas is that the farmers' old age pension is calculated independently of his former income, whereas the employees' pension is based on actual earnings.

Benefits paid under the old age pension scheme and in accident insurance are low. In 1999, the 'current pension value' in the statutory pensions insurance for workers and employees, defined as the monthly old age pension that results from contributions paid from an average income for one year, was 47.65 DM in former West Germany (40.87 DM in former East Germany). In the agricultural old age pension scheme, the current pension value is only 22.01 DM in former West Germany (18.87 DM in former East Germany). Hence, a worker with forty years of contributions and an average income will receive a pension of 1,906 DM per month, whereas a farmer with forty years of contributions will only receive 880.40 DM per month. The level of accident pension for farmers or their spouses is 36.6% of the benefit for an employee or worker with an average income, with equally reduced earning capacity. Although the farmer's accident pension is increased by 50% of his earning capacity is reduced by more than 75%, his accident pension is still only about 55% of that of an employee with an average income in a similar circumstance.

Table 1:
Differences between FRG social insurance schemes for farmers and for workers and employees[3]

	Old age pension insurance	Health insurance	Long-term care insurance	Accident insurance
Year established	1957 (1889)	1972 (1883)	1995 (1995)	1884/1939 (1884)
Insured persons	Farmers and their spouses Family members employed on the farm eligible for 50 % *(Salaried employees, wage earners)*	Full-time farmers *(Employees, workers, part-time farmers up to upper income limit)*	As for health insurance	
Eligibility criteria	Farm transfer clause Qualifying period of 180 months *(60 months)*	None	None	None
Benefits	Partial coverage Current pension value: 22.01 DM/month, 1999 (West) Additional farm aid *(Wage and contribution-related pensions Current pension value: 47.65 DM/month, 1999 (West))*	Additional farm aid *(Monetary benefits for illness)*	None	Accident rents on a low level, calculation based on a fixed farmer income of 19,440.60 DM, 1999 *(Accident rents dependent on former income and reduced earning capacity)*

[3] Regulations in the statutory systems for workers and employees and industrial sector accident insurance in parantheses and italics.

Table 1 continued

	Old age pension insurance	Health insurance	Long-term care insurance	Accident insurance
Year established	1957 (1889)	1972 (1883)	1995 (1995)	1884/1939 (1884)
Contribution system	Flat rate of contribution: 340 DM/month (old federal states), 287 DM/month (new federal states) 1999 Graduated contribution reductions from federal funds up to an income of 40,000 DM/year (Rate of contribution on earned income up to the income limit: 20.3% in 1999)	Contributions graduated according income (Rate of contribution on earned income up to the income limit, 13.6% on average old federal states in 1998)	Supplement to contributions (Rate of contribution on earned income up to the income limit, 1.7% in 1999)	Contributions graduated according to farmer's income (Contributions dependent on the sum of wages and salaries and on accident risk)
Financial assistance by federal and state grants, % of total expenditures (in 1998)	Deficit spending: 71.6% of expenditures (Federal grants: 21.7% of expenditures in 1997)	Federal grants covering the deficit of insured pensioners and their dependents: 53.6% of expenditures (–)		Yearly fixed federal grants, 34.8% of expenditures (–)

Hence, benefits for farmers provide only partial coverage and must be augmented privately.[4] This special design of the agricultural insurance system, providing only part of the social security needs of the individuals covered, is based on the assumption that farmers, like other groups of self-employed persons, have private pensions, particularly the traditional support of their family and successors.

If the farmer or his spouse becomes sick or disabled, monetary benefits will usually not be adequate to ensure that the farm can continue to be operated properly and sooner or later cease to be a reliable source of income for the family concerned. Therefore, in such cases the social security system for farmers provides farm help, i.e. substitute workers, as part of the insurance benefits to enable the continued existence of the farm.

Financing

The state's contribution to financing the social security system in agriculture varies considerably between the different schemes. In the old age pension scheme, reformed in 1995, the federal budget is responsible for deficits, i.e. it automatically pays all expenses not covered by contributions of the insured farmers and their spouses. The level of these contributions is linked to the contribution-benefit ratio of the statutory pension scheme for workers and employees. Additionally, there are income-related graduated reductions in contributions financed by federal grants (see Table 2 and later discussion).

For health and long-term care, farmers' expenditures are also largely covered by state funds. Health care, long-term care, and other benefits to retired farmers and their dependants are financed by the federal budget, not by contributions. In contrast, the health insurance systems for the nonagricultural population receives no state subsidies for retired persons; therefore employees insured in these systems pay higher contributions than do retired farmers.

---

[4] For farmers reestablishing family farms in former East Germany, however, will find it difficult have supplementary private provisions because they usually need all their financial assets to develop a viable farm enterprise. On the other hand, such persons are often entitled to pensions from the statutory old age insurance as they formerly worked on a collective or state farm.

Social Policies for German Agriculture

Table 2:
Old-age pension insurance contributions, DM/month, 1999, former West Germany

Annual income (DM) (1)	Number of insured farmers (2)	In % of insured farmers (3)	Contribution paid by farmers (4)	Contribution in statutory pension insurance* (5)	(4) in % of (5) (7)	Contribution reduction in DM/month (6)
		Farmers with graduated contribution reductions				
0–16,000	54 947	13,2	68	425	16,0	357
16,001–17,000	6 852	1,6	79	425	18,6	346
17,001–18,000	7 276	1,7	90	425	21,2	335
18,001–19,000	7 916	1,9	101	425	23,8	324
19,001–20,000	8 371	2,0	112	425	26,4	313
20,001–21,000	8 689	2,1	122	425	28,7	303
21,001–22,000	9 353	2,2	133	425	31,3	292
22,001–23,000	9 736	2,3	144	425	33,9	281
23,001–24,000	12 644	3,0	155	425	36,5	270
24,001–25,000	17 576	4,2	166	425	39,1	259
25,001–26,000	16 932	4,1	177	425	41,6	248
26,001–27,000	14 500	3,5	188	425	44,2	237
27,001–28,000	12 577	3,0	199	425	46,8	226
28,001–29,000	11 195	2,7	209	425	49,2	216
29,001–30,000	10 163	2,4	220	425	51,8	205
30,001–31,000	9 405	2,3	231	425	54,4	194
31,001–32,000	8 689	2,1	242	425	56,9	183
32,001–33,000	7 791	1,9	253	425	59,5	172
33,001–34,000	7 162	1,7	264	425	62,1	161
34,001–35,000	6 531	1,6	275	425	64,7	150
35,001–36,000	5 790	1,4	286	425	67,3	139
36,001–37,000	5 459	1,3	296	425	69,6	129
37,001–38,000	4 889	1,2	307	425	72,2	118
38,001–39,000	4 316	1,0	318	425	74,8	107
39,001–40,000	3 934	0,9	329	425	77,4	96
Total	272 693	65,4				
		Farmers without graduated contribution reductions				
40 001	143 956	34,6	340	425	80,0	85

* Pension insurance contribution of an employee for a pension at the same level as the old age pension in agriculture.
Source: Gesamtverband der Landwirtschaftlichen Alterskassen, 1999; Mehl, 1999; own calculation.

FAL-BAL

Unlike subsidisation of the other agricultural insurance schemes, federal grants to subsidise agricultural accident insurance are not legally fixed; these grants must be renewed each year in the federal government's budget.

Shortcomings

Expanding state financing In the last decade, increasing financial support to the social security system for farmers has led to a new type of income redistribution in favour of the agricultural sector in the FRG. This process was stimulated by the fact that previously the most important area of agricultural income policy, i.e. agricultural price policy, had to be left to the European Community. In addition, subsidising farmers' social security contributions was considered as a potential cushion to structural change in agriculture.

Since 1975, federal grants for subsidising social security systems in agriculture have increased considerably in relative and absolute terms (see Hagedorn, 1981b; Figure 2). In 1999, 66.6% of the budget of the German Federal Ministry for Food, Agriculture and Forestry were expenditures for social security in agriculture. In 2003, this amount will increase to approximately 72%, although the German parliament has recently taken decisions which will considerably reduce federal subsidies for all branches of social security in agriculture (Bundesministerium der Finanzen, 1999, p. 15, 21).

In the future, it will be increasingly difficult to maintain such subsidies, because the financial demands resulting from the social security systems in agriculture have their own momentum. Structural change in agriculture means that the number of people paying contributions is decreasing faster than the number of those receiving benefits. Thus the ratio between the two groups is becoming less and less favourable. The result is a disproportionate increase in the financial burden which has to be borne by each contributor.

Thus the question arises whether this additional burden is to be covered by state funds or contributions from farmers. Politicians representing farming constituencies obviously prefer the first option. However, since the budget for social security schemes in agriculture will grow more rapidly than the agricultural budget, a constant proportion of external funding of social security systems in agriculture can be maintained only if the share of the federal agricultural budget allocated to social policy rises. This erodes the funds available for other aspects of agricultural policy, e.g. improving farm structure, coastal protection. For this reason, politicians would like to reduce the amount spent on social policy in agriculture. Therefore, the choice is either to increase the costs to the agricultural budget, or to increase the burden on agricultural incomes. Both options have political drawbacks. Since the financial dynamics of the current institutional arrangement will reinforce this conflict, the best, long-term solution is to reform the system.

Figure 2:
Budget of the Federal Ministry of Food, Agriculture and Forestry and Proportion of Expenditures for Social Security Policy

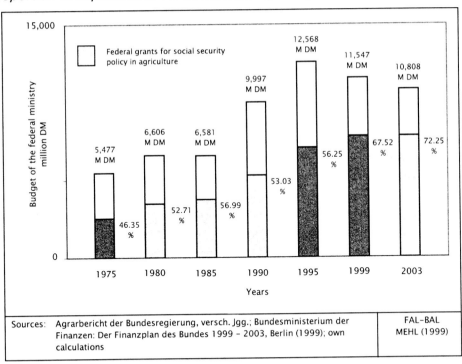

German reunification German reunification has raised the question of how to equally treat agricultural enterprises with different legal forms in social security policy (see Mehl, 1994). As already mentioned, the structure of agricultural enterprises in former East Germany, where workers and employees in large agricultural production units represent the major part of the working population on farms, differs considerably from family farms in former West Germany. While the agricultural social security system had become an important instrument of West German agricultural income policy, the question arose as to its applicability to the special situation and social security needs of the East German farm population. Introducing this special agricultural insurance system without significant changes in financing would therefore mean excluding many agricultural production cooperatives and other farms highly dependant on paid labourers (e.g. limited companies, joint stock companies), from considerable subsidies.

In searching for a solution, politicians have chosen different options in applying the social security system to agriculture in former East Germany. The West German health and accident insurance system was applied with no major changes. In the agricultural old age pension scheme, the policy transfer was combined with a partial reform of the system reducing its distributive advantages. However, as we will show, there are still considerable advantages for agriculture, even after further reform (see Mehl and Hagedorn, 1992, 1993a,b,c; 1994; Mehl, 1998). In other words, the necessity of further reforms, which should lead to a decoupling of social security policy for agriculture from income policy objectives, is reinforced not only by the increasing financial burden for the federal budget, but also by the demand for equal treatment of agricultural enterprises having different legal forms.

Reference Systems and Proposals for Reform

Criteria for Reform and a Reference System

Revision of the rules of financing is at the heart of any approach for reforming social security schemes in agriculture. Before asking what kind of alternative methods of financing might be feasible and how they should be evaluated, the principles and criteria for such a reform should be identified (Hagedorn, 1982c). The federal government has always emphasised that its main objective in offering social insurance to farmers was to guarantee them equal participation in the overall development of incomes and prosperity. This goal of equal treatment certainly characterises the basic orientation of farm politicians in the FRG, but tells us nothing about the principles and norms underlying concrete decisions. For this reason, we have to select an appropriate group whose statutory insurance schemes can serve as a frame of reference for comparison with the agricultural insurance schemes, particularly the financial and distributive impacts.

Whenever self-employed groups have been brought into statutory pension schemes, they have been subject to the same principles as employees, with some group-specific adjustments. Thus, we can conclude that social insurance for employees can be used as a frame of reference for the insurance of the self-employed. If the agricultural social security system is to provide treatment equal to that for the rest of the population, a system that truly represents the rest of the population should be used for comparison.

Thus, the statutory insurance systems for workers and employees – the statutory old age pension scheme for workers and employees,[5] their statutory health insurance,[6] and the occupational accident insurance[7] – can be regarded as the appropriate frame of reference for a reform of the social insurance schemes in agriculture, bearing in mind that this scheme is itself in need of some reform. The differences regarding the nature and level of benefits (e.g. additional support from the family or the successor, farm help in case of disability) do not affect the principle that the contribution-benefit ratio that applies to other insured persons should

---

[5] Pensions are organised in different funds: the Federal Insurance Institution for Salaried Employees in Berlin (Bundesversicherungsanstalt für Angestellte - BfA) is responsible for employees; for workers, there are various regional insurance institutions (Landesversicherungsanstalten). Migration gains accruing to the insurance funds for employees are used to contribute to the insurance funds for workers; for both organisations there is a single contribution rate on income or wages up to an income limit.

[6] Compulsory insurance covers the great majority of the working population such as employees, students, practical trainees, pensioners, unemployed persons, etc. State officials and most self-employed persons may voluntarily participate.

[7] Germany's statutory occupational accident insurance is provided by industrial, agricultural, and public-sector employers' liability insurance funds. In terms of insured persons and expenditures, the industrial insurance funds are the most important part of Germany's statutory occupational accident insurance.

also apply to farmers. This principle underlies the following comparative analysis.

An alternative approach in financing the agricultural insurance schemes in Germany faces three challenges:

- It must be based on a comparative analysis of the distributive effects, by comparing existing rules of financing social security schemes in agriculture with corresponding reference systems. This requires a more detailed analysis of the effects of the federal grants, in which two components can be distinguished (Hagedorn, 1982a). The first covers the financing deficit that arises from structural change and is attributable to the unfavourable ratio of contributors to beneficiaries. If this component were omitted, the relationship between contributions and benefits would be the same as in the statutory schemes for workers and employees. The second part of the state funding reduces the contributions of farmers to a level below the contributions workers would pay for the same pension. This subsidy component cannot be justified as compensation for structural imbalances, but is in effect an income transfer from nonagricultural sectors to agriculture.
- The programme should be practical and applicable. Requisite data should be available and administrative feasibility taken into account to create a solution ready for implementation.
- An institutional setting must be chosen that ensures social justice and political feasibility in covering the remaining structural deficits of the social security schemes in agriculture.

The Old Age Pension Scheme

To some extent principles for reform were already put into practice by a 1994 reform of the old age pension scheme for farmers. This reform had four main elements: (1) introduction of mandatory insurance of spouses of farmers with their own contributions and eligibility for benefits, (2) considerable rearrangement of benefits, (3) a new contribution system which links the development of contributions to the statutory pension insurance, simultaneously burdening the federal budget with all remaining deficits, and finally, (4) elements to make such reform politically feasible, i.e. the existing graduated reductions in contributions are stepwise lowered until 2010, and this process is additionally cushioned by transitional provisions, to avoid abrupt decreases in benefits.

These changes will have considerable effects. Introduction of mandatory insurance for the spouses of farmers was not only an important progress regarding social security for this group. It also helped stabilise the system in the short run by increasing the number of contributors.

In the long run, the system will be stabilised by cuts in benefits. A supplement for married farmers is no longer paid; it has been replaced by the spouse's insurance. Before 1995, old age pensions in agriculture were graduated according to how many years contributions were made by means of a very special regulation. After fifteen years of contribution the farmers were entitled to a basic pension, for each additional year of contribution it was increased by 3 %. In other words, the first fifteen years were credited with 6.7% of the basic pension per year (100% for fifteen years of contributions), and the following years with 3% per year. This meant that the high basic pension resulted in a redistribution of income to farmers who had paid contributions for only a short period of time. After the reform, the calculation of benefits takes account of each year equally, as in the German statutory pension scheme for workers and employees. In some cases the new regulation leads to a drastic reduction of benefits. With the old regulation, a married farmer with fifteen years of contributions would have received 754.13 DM/month (figures for 1998, former West Germany); under the new system he will only receive 330.15 DM/month, but his spouse will be entitled to benefits due to own contribution payments (see Mehl, 1999, p. 146).

As already mentioned, contributions to the agricultural old age pension scheme are linked to the contribution–benefit ratio of the statutory pension scheme: A farmer pays 80% of a worker's contribution to receive an equal pension (his 'standard contribution'); the federal government assumed that the 20% reduction in contributions was justified by fewer benefits offered than under the workers' system. In other words, by guaranteeing farmers and their spouses a contribution–benefit ratio equal to the statutory pension insurance, the federal government has agreed to bear the financial consequences of structural change in agriculture, as far as it affects old age insurance.

However, only about 50% of insured farmers have to pay the standard contribution due to a further relief of contributions by federal subsidies used for graduated contribution reductions. Farmers and their spouses fulfilling the conditions for these graduated reductions are entitled to obtain additional subsidies that reduce the standard contribution by up to 80%.[8]

Compared with the distributive effects the system produced before it was reformed, a considerable decrease in income subsidisation has taken place. However, the financial arrangements still provide considerable advantages for the insured farmers and their spouses. As mentioned, the changes are implemented gradually during a transition period which will last until 2010. This strategy of a 'soft transition' was based on the argument that insured persons would be limited in their ability to adjust quickly to changing conditions. Additionally, those farmers eligible for graduated contribution reductions still draw considerable advantages from membership in the agricultural system. In 1999, 64.5% of all insured farmers were entitled to graduated contribution reductions, and only 35.5% had to pay the standard contribution (340 DM per month), which is only 80% of the contribution required of those under the statutory pension insurance (425 DM per month) for the same pension. About 13% of the insured farmers are eligible for the highest reductions in contributions; they pay a monthly contribution of 68 DM, only 16% of the contribution an employee in the statutory pension insurance is charged for the same pension (Table 2).

Such a combination of an old age pension independent of income with an income-dependent contribution creates many redistribution and administration problems for which as yet, there is no satisfactory solution (for details, see Hagedorn, 1987; Mehl, 1996). Above all, a conflict arises in that provisions for old age, death of the breadwinner, illness, accident, and so forth cannot concomitantly be used as an instrument for transferring income to agriculture in a reasonable way. Since social security contributions should be calculated individually according to income, reductions in contributions are not compatible with the principles of social income policy: the lower the income, the more needy the individual is, but also the lower are his contributions. Consequently, the

---

[8] Farmers with incomes up to 40,000 DM per year (80,000 per year for a farmer and spouse) are eligible for further reductions; the highest graduated reductions are paid to farmers with incomes of 16,000 DM per year or less (32,000 per year, farmer and spouse).

increase in income which can be achieved by reducing or waiving his contribution is smaller, too. Thus, reductions in contributions are an inefficient income policy strategy.

Health Insurance

Amendments that cut expenditures to the general statutory health insurance in the late 1980s and 1990s were also made to agricultural health insurance, since the design and criteria of benefits are almost identical and benefit cuts in the general system are automatically implemented in the agricultural system. Therefore, to realise further savings, the focus must be on aspects of financing particular to agricultural health insurance.

Since 1972, federal grants have financed expenditures for pensioners and their dependants in the agricultural health insurance system that are not covered by participants' contributions (Hagedorn, 1982b). In the statutory health insurance for workers and employees, contributions of pensioners are also insufficient to cover expenses. In this case, however, the deficit must be covered by contributions from insured workers and employees, because federal grants are unavailable. Since 1972, the proportion of contributions needed to finance health insurance costs for pensioners in the general statutory system has increased considerably (Table 3). In 1973, 10.57% of the contributions by employees and workers were spent for this purpose; by 1995, approximately 30%. The contribution rate to the statutory health insurance programme went from 9% of wages and salaries in 1973 to 13.18% in 1995. If the (favourable) regulations of the agricultural health insurance were applied to the statutory programme, the average contribution rate would only have increased from 8.05% (1973) to 9.23% (1995). Hence, the distributive advantages in the agricultural health insurance programme have grown considerably since 1973. Since the contribution scheme of the agricultural health insurance and the agricultural long-term care insurance are connected, these advantages were transferred to the latter system, too.

We see the following possibilities for equalising the two systems. Simply limiting the federal grants that cover the deficit expenses for agricultural pensioners to a certain percentage seems unreasonable. Such a limit would either underestimate or overestimate the burden on the agricultural system caused by structural change.

The statutory health insurance in the FRG consists of different institutions, that have varying geographic or occupational scope. Since 1996, however, wage earners as well as salaried employees may choose between various funds. To avoid dispersing contributions, a compensation system for unequal risk structures ('Risikostrukturausgleich', or RSA) was introduced (Wasem, 1993; Minnich and Pfeiffer, 1995). This RSA consists of compensation payments from insurance institutions that are better off in favour of institutions with disadvantageous risk structures. Hence, the system is in accordance with the aim to ensure equal treatment of insured farmers under the agricultural health insurance. Therefore it seems quite reasonable to integrate agricultural health insurance into this system of transfers for determining the amount of federal grants in financing the agricultural health insurance. Unfortunately, including agriculture in the RSA system would require some data what is unavailable yet. Considerable field research is needed.

A reasonable answer to the question of how much active farmers should contribute to financing the deficit arising from health insurance for retired farmers is by defining a 'supplement to their contribution'. The level of this supplement should be determined by calculating the burden that employees and workers under statutory health insurance have to accept for financing deficits for their pensioners (Table 3). In 1995, 29.96% of the contributions of those workers and employees were applied for this purpose, increasing the contribution rate from 9.23% of earnings and salaries to 13.18% in 1995. Applying this formula to farmers in 1995, the supplement would have totalled 456.77 million DM, equal to 22.94% of the federal grants to agricultural health insurance that same year.

Accident Insurance

Despite the amendment of the statutory accident insurance in 1997, accident pensions for farmers in the agricultural accident insurance remained low to avoid increasing their contributions. Hence, the only opportunity to lower expenditures in this system is by reducing benefits. Therefore, the financing of agricultural accident insurance must be scrutinised. Unlike the federal grants for the agricultural old age pension

Table 3:
Deficits in Statutory Health Insurance for Pensioners and Corresponding Contribution Supplements for Farmers Insured in the Agricultural Health Insurance, 1973 – 1995

Year	Deficits in the statutory health insurance for pensioners (KVdR) (1) Mio. DM	Deficits of the KVdR in % of contributors (2) in %	Average contribution rate in the statutory health insurance (3) in %	Part of contributions spent for covering deficits of the KVdR (4) in %	Contribution without deficit covering for pensioners (5) in %	Contributions of farmers in the agricultural health insurance (LKV) (6) Mio. DM	Contribution supplement in the LKV to joint financing of the deficits of pensioners in the LKV (2) x (6) (7) Mio. DM	Contribution of farmers + supplement (6) + (7) (8) Mio. DM	Federal grants for financing the deficit of pensioners in the LKV (9) Mio. DM	Contribution supplement in % of federal grants (10) in %
1973	3.506	10,57	9,00	0,95	8,05	621,30	65,66	686,96	430,00	15,27
1974	4.834	12,82	9,47	1,21	8,26	741,40	95,03	836,43	520,00	18,28
1975	5.279	11,74	10,04	1,18	8,86	856,00	100,52	956,52	630,00	15,96
1976	5.066	9,76	10,53	1,03	9,50	907,40	88,56	995,96	662,80	13,36
1977	7.779	13,91	11,22	1,56	9,66	913,10	127,01	1.040,11	688,00	18,46
1978	10.892	18,19	11,30	2,05	9,25	956,00	173,85	1.129,85	723,90	24,02
1979	12.781	20,14	11,40	2,30	9,10	976,60	196,65	1.173,25	798,50	24,63
1980	15.080	21,81	11,37	2,48	8,89	1.032,60	225,17	1.257,77	881,30	25,55
1981	17.449	23,02	11,75	2,70	9,05	1.058,40	243,65	1.302,05	986,60	24,70
1982	17.841	22,19	11,98	2,66	9,32	1.081,30	239,98	1.321,28	996,50	24,08
1983	20.142	24,59	11,93	2,93	9,00	1.167,20	286,96	1.454,16	951,10	30,17
1984	22.472	27,13	11,46	3,11	8,35	1.196,00	324,51	1.520,51	1.009,50	32,15
1985	25.015	28,22	11,73	3,31	8,42	1.293,70	365,10	1.658,80	1.067,60	34,20
1986	26.985	28,46	12,15	3,46	8,69	1.317,00	374,79	1.691,79	1.147,60	32,66
1987	28.662	28,45	12,47	3,55	8,92	1.324,10	376,71	1.700,81	1.191,20	31,62
1988	31.865	29,70	12,89	3,83	9,06	1.318,70	391,61	1.710,31	1.257,60	31,14
1989	27.286	24,50	12,78	3,16	9,74	1.315,50	322,29	1.637,79	1.284,40	25,09
1990	29.719	25,53	12,22	3,26	9,52	1.299,20	331,68	1.630,88	1.360,80	24,37
1991	37.169	30,51	12,22	3,73	8,49	1.327,00	404,81	1.731,81	1.577,40	25,66
1992	43.980	32,57	12,46	4,06	8,40	1.410,50	459,41	1.869,91	1.821,70	25,22
1993	40.622	27,36	13,42	3,67	9,75	1.504,50	411,57	1.916,07	1.795,70	22,92
1994	43.530	28,68	13,35	3,83	9,52	1.475,60	423,21	1.898,81	1.908,70	22,17
1995	46.668	29,96	13,18	3,95	9,23	1.524,40	456,77	1.981,17	1.990,90	22,94

Source: Bundesministerium für Arbeit und Sozialordnung, 1982; 1987; 1997; Mehl, 1999; own calculations.

FAL–BAL
MEHL (1999)

scheme and agricultural health insurance, the amount of federal grants to agricultural accident insurance is fixed in the federal budget and must be approved each year. Consequently the federal contribution has fluctuated, something that has been criticised on various occasions (Wissenschaftlicher Beirat, 1979; von Maydell, 1988). A method developed (or adopted from other sectors) to fix the amount of federal grants for agricultural accident insurance must assure that insured farmers need not bear an excessive burden caused by structural change, and grants which only aim at subsidising farmer incomes should be withdrawn.

It may be useful to learn from the thirty-five German accident insurance organisation in the industrial and trade sector which are organised according to branches of the economy. In order to achieve a balance between trade and industry, a compensatory system was established in 1968 (see Schimanski, 1986). The system is built on fixed mathematical rules, meaning that entitlement to compensatory payments and the obligation to make compensatory payments are set every year under the same principles. Measurement of an excessive burden is based on the ratio between accident pensions expenditures and the sum of earnings and salaries subject to making contributions. An excessive burden exists if the rent burden ratio (Rentenlastsatz) of an insurance organisation exceeds the average ratio of all accident insurance organisations in trade and industry by more than 4.5 times. This compensation mechanism in the statutory accident insurance in trade and industry has worked exclusively in favour of the mining and inland navigation sectors, both of which are experiencing structural decline.

In particular the situation of the accident insurance organisation for mining seems comparable with the situation of agricultural accident insurance. Both suffer from a considerable decrease in the workforce while expenditures for beneficiaries of accident insurance continue to rise. Hence, the method described above may also be suitable to clarify whether and to what extent an excessive burden exists in the agricultural branch of the statutory accident insurance. As the administrative procedure for collecting contributions is arranged differently in the agricultural insurance scheme, the sum of earnings and salaries is not available. Therefore, the numerator of the rent burden ratio has to be estimated by multiplying the figure of working units (Vollarbeiter) by the so-called 'fixed average income' of farmers in agricultural accident insurance. However, since the latter is only an artificial reference point exclusively used in accident pension calculation, and does not reflect actual incomes, this approach implies an overestimation of the rent burden ratio in agricultural accident insurance (Scheele, 1989a; 1989b).[9]

Figure 3:
Burden Ratio of Accident Pensions in the Industrial and Agricultural Accident Insurance and Power Limits for Compensation Payments According to the Rules of Burden Sharing in the Industrial Accident Insurance in Germany 1963 - 1997

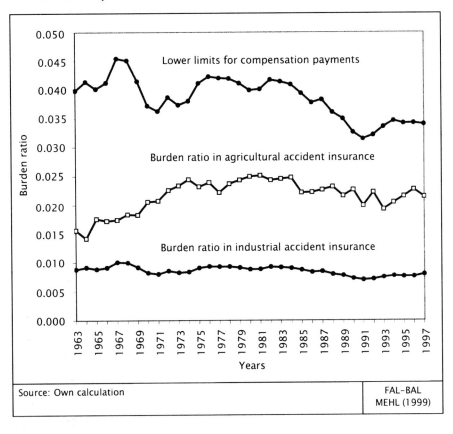

As a consequence, transferring the calculation method from industrial accident insurance organisations to accident insurance in agriculture

---

[9] This approach especially underestimates the income relevant to determine the contributions of workers and employees in the agricultural sector which are insured in agricultural accident insurance.

would be misleading: it would only give the impression that the burden of accident pensions in the agricultural system had been considerably higher than that in the industrial branches between 1963 and 1997. In those years, however, the threshold at which compensation payments would be made for lowering an excessive burden was not reached. Accordingly, agricultural accident insurance would not be eligible for compensatory payments if the nonagricultural rules outlined above were even partially applied (see Figure 3).

However, these calculations reveal nothing about the burden ratio in the different regional organisations of agricultural accident insurance, because reliable data are not yet available on the regional level. If we consider the quite different regional structures in German agri-culture, presumably some regional institutions are likely to reach the threshold of excessive burden and would therefore be eligible for compensation payments. However, according to the principles mentioned above, compensation between regions is not the responsibility of federal grants, but of the regional organisations in agricultural accident insurance as a matter of solidarity.

Alternative Ways of Financing the Deficits of Social Security Systems in Agriculture

Probably the first-best solution for covering deficits in the agricultural insurance schemes due to structural change would be to introduce migration compensation agreements by which losses caused by people leaving the agricultural insurance schemes would be compensated for by the gains of the statutory schemes (see Hagedorn, 1987; 1991). This would be an important first step in a process aimed at creating an integrated financing system, which might have the effect of a 'constitutional constraint' to the budget, increasing the 'transaction costs' to using the social security system for rent-seeking objectives. However, such a solution hardly seems feasible due to administrative and political obstacles, including:

- accurate monitoring of migration gains from other social security institutions in the FRG, e.g. the systems for state officials or systems for self-employed professions

- obtaining reliable data needed to integrate agricultural health insurance into the mechanism of burden sharing necessary to cope with unfavourable risk structures, particularly reliable information on the incomes of insured farmers
- the additional rent burden caused by transitional regulations concerning old age pension benefits, which up to 2010 give farm pensioners a much better contribution–benefit ratio than does the statutory pension insurance
- resistance by trade unions and employers' associations who want to avoid the additional burden hitherto covered by grants from the federal budget

Therefore, it might be reasonable to limit the federal financing of the agricultural schemes to the point at which the contribution–benefit ratio is equivalent to the social security schemes of employees and workers. If the rules for financing the system were rearranged in this way, savings could be used to finance programmes better suited to assisting working and retired farmers in solving income and adjustment problems (see Hagedorn, 1988).

The Future of Social Security Systems for Farmers in Germany

Unless further reforms are made, the FRG's agricultural insurance system will become even more overstretched. It will become increasingly dependent on state subsidies because its financial deficit increases for macroeconomic and demographic reasons (see Hagedorn, 1986b; Mehl, 1995). In the process of structural change, the agricultural pension scheme loses contributors to other schemes, and these schemes gain contributors from the agricultural sector. Therefore, an appropriate way of covering it would transfer contributions lost to other schemes back to the agricultural system.

Financing the structure-induced revenue gap in this way should not be combined with income transfers that further reduce contributions, since such transfers cannot be properly targeted within the agricultural social insurance scheme. Attempting to solve income problems by reducing contributions takes no account of individual hardship. This is because reductions offer insufficient flexibility when income support is to be targeted at low income earners. A reformed financing system must go no further than compensating for the structure-induced deficit. This constraint must be embodied in institutional barriers to prevent the use of general reductions in contributions to make undifferentiated income transfers.

The structural deficit should therefore no longer be covered by federal funds, but by the gains in nonagricultural pension schemes. The main reason for this is that it clarifies the transfer system and prevents rent-seeking activity by agricultural pressure groups (Hagedorn, 1994; 1996; 1998). The most appropriate organisational means to do this would be a financing institution covering old age insurance for both farmer and employees and workers, as already in operation for old age insurance for employees and workers. In the past, total integration with the workers' and employees' schemes was often considered inadvisable, because of political resistance to terminating the organisational independence of the agricultural social security system. However, since the working population in agriculture has become very small, much of it already in nonagricultural insurance organisations due to the effects of German reunification, complete integration may be the most practical solution. This seems to be even more advisable since the traditional objectives that were used to justify a special social security system for farmers (e.g. the farm transfer clause as a means to accelerate structural change) are less relevant.

In the future, some functions which belonged to the domain of the social security system for farmers, will probably be gradually shifted into the domain of environmental policies. First, contributions to the farmers' social security organisations will be partly replaced by ecological taxes levied by the tax offices. In Germany, ecological tax reform, in several steps, is planned. The first step was finalised in 1999; the agricultural economy (i.e. farm enterprises) is taxed 343 million DM per year, and the agricultural population (i.e. farm-related households), 128 million DM per year. Simultaneously, social security contributions are partly replaced by the revenue from ecological taxes, enabling the old age insurance administration to demand 40 million DM per year less from self-employed farmers insured by the agricultural old age pension scheme, 58 million DM per year less from farm workers insured in the statutory pension insurance. This shows that agriculture does not profit financially from the ecological tax reform, and also that it is questionable whether the so-called 'double dividend', the desired combination of positive ecological and employment effects, will have a sizeable impact (for details, see Hagedorn, 1999).

Secondly, social policy for agriculture has more or less lost its political function. The expansion of social security policies in Western European countries can be traced back to a political era when the welfare state had a very positive image. Given the political climate at the time, it was reasonable for farm politicians to embody their political intentions in social programmes to legitimise and garner support for their group-specific interests (for empirical evidence, see Hagedorn, 1993). Combining agricultural income policies with social security schemes was a rational strategy for political entrepreneurs. Today most of the large social security systems in developed countries are in considerable difficulty and require fundamental reform that demands sacrifice from their members. Since this is also true for the social security schemes in agriculture, they have become less instrumental in promoting agricultural interests. Instead, ecologically oriented policies have assumed the role of legitimising agricultural policies. The most recent example is the Agenda 2000 which tries to link transfer payments which have arisen from the McSharry Reform to ecological standards, without reliable knowledge as to whether this type of 'cross compliance' will be politically feasible and ecologically effective (see Eggers and Hagedorn, 1998). In other words, the task of providing legitimisation for agricultural interests is obviously shifting from social to environmental policies.

References

Agrarbericht der Bundesregierung (various years). Bonn.

Beckmann, V. (2000). Transaktionskosten und institutionelle Wahl in der Landwirtschaft. Zwischen Markt, Hierarchie und Kooperation. Berliner Studien zur Kooperationsforschung, Berlin Co-operative Studies 5. Berlin: edition sigma.

Bundesministerium der Finanzen (1999). Der Finanzplan des Bundes 1999–2003. Bonn.

Bundesministerium für Arbeit und Sozialordnung (various years). Arbeits- und Sozialstatistik. Bonn.

Bundesministerium für Arbeit und Sozialordnung (various years). Sozialbericht. Bonn.

Bundesministerium für Arbeit und Sozialordnung (1997). Arbeitssicherheit 1997. Bericht der Bundesregierung über den Stand der Unfallverhütung und das Unfallgeschehen in der Bundesrepublik Deutschland im Jahr 1996. Bonn.

Deutsche Bundesbank (1996). Zur Diskussion über die öffentlichen Transfers im Gefolge der Wiedervereinigung. Deutsche Bundesbank - Monatsbericht 48, No. 10: 17-31.

Deutscher Bundestag (1999). Antwort der Bundesregierung auf die kleine Anfrage der CDU/CSU-Bundestagsfraktion: Sicherung des agrarsozialen Systems. Bundestagsdrucksache 14/1382 vom 7.7.1999.

Eggers, K. J. (1980). Agrarsoziale Sicherung im EG-Vergleich. Münster-Hiltrup: Landwirtschaftsverlag.

Eggers, J. and Hagedorn, K. (1998). Umwelteffekte und agrarumweltpolitische Ansätze der "Agenda 2000". Agrarwirtschaft 47, No. 12: 482-491.

Fink, M., Grajewski, R., Siebert, R. and Zierold, K. (1994). Rural Women in East Germany. In: Symes, D.; Jansen, A. J., (eds.) Agricultural Restructuring and Rural Change in Europe, Wageningen: Agricultural University 1994: 282-295.

Forstner, B. and Isermeyer, F. (2000). Transformation of Agriculture in East Germany. In this volume.

Gesetz zur Reform der agrarsozialen Sicherung (1995). Agrarsozialreformgesetz 1995 - ARSG 1995. BGBl., 1890.

Hagedorn, K. (1981a). Die Hofabgabeklausel in der Altershilfe für Landwirte. Agrarrecht 11, No. 1: 7-10.

Hagedorn, K. (1981b). Voraussichtliche Entwicklung der Finanzierung der landwirtschaftlichen Alterssicherung. Vierteljahresschrift für Sozialrecht 9, No. 1/2: 111-147.

Hagedorn, K. (1982a). Agrarsozialpolitik in der Bundesrepublik Deutschland. Kritik und Alternativmodelle zur Alterssicherung in der Landwirtschaft. Beiträge zur Sozialpolitik und zum Sozialrecht, Vol. 1. Berlin: E. Schmidt.

Hagedorn, K. (1982b). Die Finanzierung der Krankenversicherung für landwirtschaftliche Altenteiler. Agrarwirtschaft 31: 165-172.

Hagedorn, K. (1982c). Ein Konzept zur Neugestaltung der landwirtschaftlichen Alterssicherung. Landbauforschung Völkenrode 32, No. 1: 43-56.

Hagedorn, K. (1986a). Reformversuche in der Geschichte der Agrarsozialpolitik. Zeitschrift für Agrargeschichte und Agrarsoziologie 34: 176-215.

Hagedorn, K. (1986b). Ökonomische und politische Auswirkungen der rückläufigen Bevölkerungsentwicklung auf die Finanzierung der agrarsozialen Sicherung. In: von Blanckenburg P. and de Haen, H. (eds.) Bevölkerungsentwicklung, Agrarstruktur und Ländlicher Raum. Schriften der Gesellschaft für Wirtschafts-

und Sozialwissenschaften des Landbaues e. V., Vol. 22. Münster-Hiltrup: Landwirtschaftsverlag: 259-272.

Hagedorn, K. (1987). Alternative Modelle zur Finanzierung der landwirtschaftlichen Alterssicherung. Landbauforschung Völkenrode 36: 249-267.

Hagedorn, K. (1988). Political Economics in Social Programmes for Resource Adjustment in EC-Agriculture. Forum, 18. Kiel: Vauk.

Hagedorn, K. (1991). Financing Social Security in Agriculture: The case of the farmers' old-age pension scheme in the Federal Republic of Germany. European Review of Agricultural Economics 18: 209-229.

Hagedorn, K. (1993). Umweltpolitische und sozialpolitische Reformen in der Agrarpolitik. Parallelen und Unterschiede zwischen phasenverschobenen Politikprozessen. Zeitschrift für Umweltpolitik und Umweltrecht 16, No. 3: 235-280.

Hagedorn, K. (1994). Interest Groups. In: Hodgson, G., Tool, M. and Samuels, W. J. (eds.). Handbook of Institutional and Evolutionary Economics. Cheltenham: Edward Elgar: 412-418.

Hagedorn, K. (1996). Das Institutionenproblem in der agrarökonomischen Politikforschung. Tübingen: Mohr.

Hagedorn, K. (1997). The Politics of Privatization of Nationalized Land in Eastern Germany. In Swinnen, J. F. M. (ed.). Political Economy of Agrarian Reform in Central and Eastern Europe. Aldershot, Brookfield USA, Singapore, Sidney: Ashgate: 197-236.

Hagedorn, K. (1998). Reasons and Options for Analyzing Political Institutions and Processes. In Frohberg, K. and Weingarten, P. (eds.). The Significance of Politics and Institutions for the Design and Formation of Agricultural Policy. Studies about the Agri-Food Sector in Central and Eastern Europe, Vol. 2. Edt. by the Institut für Agrarentwicklung in Mittel- und Osteuropa (IAMO). Kiel: Vauk. 14-33.

Hagedorn, K. (1999). Der Double Dividend-Ansatz als institutionelle Weichenstellung zur Ökologisierung der Agrarpolitik? In: Mehl, P. (ed.). Agrarstruktur und ländliche Räume. Festschrift zum 65. Geburtstag von Eckhart Neander. Braunschweig 1999. Landbauforschung Völkenrode, Sonderheft 201: 177-210.

Hagedorn, K. and Beckmann, V. (1995). De-collectivisation Policies and Structural Changes of Agriculture in Eastern Germany. MOCT-MOST 5, No. 4: 133-152.

Hagedorn, K. and Beckmann, V. (1997). De-collectivisation Policies and Structural Changes of Agriculture in Eastern Germany. In Swinnen, J. F. M., Buckwell, A.

and Mathijs, E. (eds.). Agricultural Privatisation, Land Reform and Farm Restructuring in Central and Eastern Europe. Aldershot, Brookfield USA, Singapore, Sidney: Ashgate: 105-160.

Maydell, B. Baron von (1988). Weiterentwicklung des landwirtschaftlichen Sozialrechts. Möglichkeiten der Weiterentwicklung des landwirtschaftlichen Sozialrechts unter besonderer Berücksichtigung der rechtlichen Rahmenbedingungen. Schriftenreihe des Bundesministeriums für Ernährung, Landwirtschaft und Forsten, Reihe A: Angewandte Wissenschaft, No. 352. Münster-Hiltrup: Landwirtschaftsverlag.

Mehl, P. (1994). Sozialrechtliche Behandlung unterschiedlicher Rechtsformen. In: Klare, K. (ed.). Entwicklung der ländlichen Räume und der Agrarwirtschaft in den Neuen Bundesländern. FAL-Symposium am 2. und 3. November 1994. Schriftliche Fassung der Beiträge. Landbauforschung Völkenrode 44, Sonderheft 152: 91-103.

Mehl, P. (1995). Demographische Strukturen von morgen und ihre Wirkungen auf die Funktionsfähigkeit der agrarsozialen Sicherung. In: Agrarsoziale Gesellschaft e. V. (Hrsg.). Anforderungen an die Gesellschaft vor dem Hintergrund demographischer Veränderungen. Schriftenreihe für ländliche Sozialfragen, No. 120. Göttingen: 67-79.

Mehl, P. (1996). Die Mitwirkungspflicht der Versicherten bei der Feststellung einer Beitragszuschußberechtigung im Rahmen der landwirtschaftlichen Alterssicherung. Stellenwert und Sanktionierung etwaiger Verstöße. Schriftliche Stellungnahme aus Anlaß der öffentlichen Anhörung des Ausschusses für Arbeit und Sozialordnung des Deutschen Bundestages zum Gesetzentwurf der Fraktionen der CDU/CSU und FDP: Entwurf eines Gesetzes zur Änderung des Gesetzes über die Alterssicherung der Landwirte. Bundestags - Drucksache 13/4847 am 6.11.1996 in Bonn. Ausschuß für Arbeit und Sozialordnung des Deutschen Bundestages, Ausschuß-Drucksache 851

Mehl, P. (1997a). Reformansätze und Reformwiderstände in der Agrarsozialpolitik der Bundesrepublik Deutschland. Politikinhalte und Ihre Bestimmungsgründe von 1976 bis 1990. Sozialpolitische Schriften, Berlin: Duncker & Humblot.

Mehl, P. (1997b). Transforming Social Security in Agriculture in Transition Countries: The Case of East Germany. Landbauforschung Völkenrode 47: 75-89.

Mehl, P. (1998). Transformation of the Social Security System in Agriculture in East Germany. Lessons for Central and Eastern European Countries? In: Frohberg, K. and Weingarten, P. (eds.). The Significance of Politics and Institutions for the

Design and Formation of Agricultural Policies. Studies of the Agricultural and Food Sector in Central and Eastern Europe, Vol. 2. Kiel: Vauk: 139-156.

Mehl, P. (1999). Agrarsoziale Sicherung als politisches Reformproblem. In: Mehl, P. (ed.). Agrarstruktur und ländliche Räume. Festschrift zum 65. Geburtstag von Eckhart Neander. Landbauforschung Braunschweig-Völkenrode, Sonderheft 201: 129-162.

Mehl, P. and Hagedorn, K. (1992). Übertragung der agrarsozialen Sicherung auf die neuen Bundesländer: Probleme und Perspektiven. Landbauforschung Völkenrode 42: 276-292.

Mehl, P. and Hagedorn, K. (1993a). Agrarsoziale Sicherung in den Neuen Bundesländern. In: Alvensleben, R. von; Langbehn, C. and Schinke, E. (eds.). Strukturanpassung in der Land- und Ernährungswirtschaft in Mittel- und Osteuropa. Schriften der Gesellschaft für Wirtschafts- und Sozialwissenschaften des Landbaues e. V., Vol. 29. Münster-Hiltrup: Landwirtschaftsverlag: 311-320.

Mehl, P. and Hagedorn, K. (1993b). Eigenständige soziale Sicherung der Bäuerin und finanzielle Stabilisierung des agrarsozialen Sicherungssystems. Überlegungen zum Gesetzentwurf der Bundesregierung zur Reform der agrarsozialen Sicherung (ASRG 1995). Agra-Europe 34, Nr. 49, Sonderbeilage: 1-9.

Mehl, P. and Hagedorn, K. (1993c). Die agrarsoziale Sicherung im Prozeß der Vereinigung Deutschlands - Probleme des Übergangs zu einem sektoral gegliederten Sozialversicherungssystem. Deutsche Rentenversicherung o. Jg., H. 3: 120-147.

Mehl, P. and Hagedorn, K. (1994). Die Übertragung des landwirtschaftlichen Alterssicherungssystems auf die Neuen Bundesländer im Gesetzentwurf der Bundesregierung zur Reform des agrarsozialen Sicherungssystems. Landbauforschung Völkenrode 44: 77-90.

Minnich, N. and Pfeiffer, D. (1995). Der Risikostrukturausgleich - Solidarität in einer wettbewerblich orientierten Krankenversicherung. Die Ersatzkasse 75: 121-126.

Scheele, M. (1989a). Sektorale Einkommensübertragungen im Rahmen der Agrarsozialpolitik. Agrarwirtschaft 38: 203-213.

Scheele, M. (1989b). Zur Ermittlung des strukturwandelbedingten Defizits in der landwirtschaftlichen Unfallversicherung. Agrarwirtschaft 38: 350-351.

Schimanski, S. (1986). Risiko- und Solidaritätsausgleich in der gesetzlichen Unfallversicherung. Eine Untersuchung zur Methode, Durchführung und Wirksamkeit des Lastenausgleichsverfahrens nach Art. 3 UVNG in der Fassung des Finanzänderungsgesetzes vom 21.12.1967. Dissertation. Bochum.

Tegtmeier, W. (1993). The Unification of the Two German Social Systems: Problems and Possible Consequences for the Conversion Process in Central and Eastern Europe. In: European Institute of Social Security (EISS) (1993). Problems of Transformation of Social Protection Systems in Central and Eastern Europe. Yearbook 1993, Leuven/Amersfoort: 51-81.

top agrar Spezial (1992). No. 3: 41 (Neue Bundesländer: Abwanderung aus der Landwirtschaft).

Wasem, J. (1993). Der kassenartübergreifende Risikostrukturausgleich. Chancen für eine neue Wettbewerbsordnung in der GKV. Sozialer Fortschritt, 42: 32-29.

Winkler, W. (1992). Die Altersversicherung der Landwirte in der Europäischen Gemeinschaft. Soziale Sicherheit in der Landwirtschaft. No. 2: 214-248.

Wirth, C. (1994). Reform der Alterssicherung der Landwirte. Soziale Sicherheit in der Landwirtschaft. No. 2: 67-100.

Wissenschaftlicher Beirat beim Bundesministerium für Ernährung, Landwirtschaft und Forsten (1979). Agrarsozialpolitik – Situation und Reformvorschläge. Münster-Hiltrup: Landwirtschaftsverlag.

Chapter 8

Institutional Characteristics of German Agriculture

Werner Großkopf
Institute of Agricultural Policy and Market Research
University of Hohenheim-Stuttgart

Introduction

There are several branches of economic theory which concern themselves with the economic analysis of institutions. This chapter will deal with business structures (institutions), in which transaction costs play an important role. Transaction costs consist of three major components: the costs for information, negotiation, and supervision. Transaction cost theory investigates the conditions leading to and surrounding the creation of institutions, their contribution to the reduction of transaction costs, and the level of transaction costs within different institutions. Farms have banded together to form their own institutions, both with and without state assistance, in order to lower their transaction costs and benefit from economies of scale, and to strengthen their market power. These institutions can be classified into four types (see also Table 1):

- Institutions aimed at reducing transaction costs in the political field, i.e. institutions that exert influence on agricultural policy. Institutions carrying political influence include those that represent the interests of the agricultural sector. The most important of these are the Deutscher Bauernverband e.V. (German Farmers' Association or DBV), the Deutscher Raiffeisenverband e.V. (German Raiffeisen Association or DRV), the Deutsche Landwirtschafts-gesellschaft e.V. (the German Agricultural Society or DLG), and the

Verband der Landwirtschaftskammern e.V. (the Association of the Chambers of Agriculture).
- Institutions that conduct information, counselling, and training functions. This group consists of the chambers of agriculture, amongst others, which are autonomous, professional, self-governing institutions. As corporations under public law they function as self-governing institutions and are self-administrating (Henrichsmeyer and Witzke, 1994, p. 431). Furthermore, they perform an organisational role for the providers of counselling and information services.
- Institutions focusing on the cooperation of farms in the joint use of the means of production in farming. These ring organisations have developed into diverse forms over the years.
- The final group consists of institutions whose function is to foster cooperation in procurement, sales, and marketing for farms. The rural cooperatives represent important self-help institutions for the agricultural sector on the agricultural commodities markets.

Table 1:

Categories of institution groups in Germany

	Institutions exerting influence	Decision-making institutions	Cooperation in production	Cooperation in marketing
Government aid	−	+	+	−
Reduction of transaction costs	+	−	−	+
Use of economies of scale	−	−	+	+
Strengthening of market power	+	−	−	+

+ present
− not present, or only to a minimal extent

Institutional Characteristics of German Agriculture

Representatives of German Agricultural Interests

Agricultural Interest Groups

Different trade organisations present German agricultural interests to the public and, above all, to the decision-makers of agricultural policy. Together the four 'umbrella' agricultural interest groups form the Zentralausschuß der deutschen Landwirtschaft (Central Committee of German Agriculture or ZDL), founded in 1949. These are the four 'umbrella organisations' of German agriculture: the DBV, the DRV, the DLG, and the Association of the Chambers of Agriculture (Henrichsmeyer and Witzke, 1994, p. 434). The political control of the ZDL clearly lies in the hands of the DBV (Maisack, 1995, p. 10). By effectively stapling together its member organisations, the DBV operates 'a type of representative monopoly' for German agriculture.

The German Farmers' Association as Representative of German Agriculture

The origins of the representation of agricultural interests to the ruling powers in Germany lie in the 18th century when agricultural associations were developed by leading farmers, although a broad-based movement did not begin until 1862. When the National Socialists rose to power, the agricultural associations were either dissolved or assimilated into the Reichsnährstand, the universal agricultural organisation of the Reich. Membership in this organisation was compulsory for all farmers and processors of agricultural products, as well as for trade, marketing and professional associations. After the Reichsnährstand was broken up by the Allies in 1948, the organisational vacuum was filled by the state farmers' associations, which were founded in 1945 (Mändle, 1983, p. 73). In 1948, the farmers' associations in the three West German zones were amalgamated to form the DBV. The driving force for this was the desire for a general, united association that was both politically and denominationally independent. This association was to represent the interests of all sectors of production, plant sizes, and property types on a voluntary and democratic basis. Up to the time of reunification, the agricultural sector in the Soviet-occupied zone went its own way (Schnieders, 1998, p. 4-10).

The DBV sees itself as 'the professional representative of people working in the sectors of agriculture and forestry, as well as related industries, in the Federal Republic of Germany'. According to its charter, its task is to represent 'the agricultural, legal, tax, welfare, educational and social interests' of the agricultural and forestry sectors and 'to co-ordinate the activities of the member organisations in all essential affairs' (Deutscher Bauernverband, 1998, p.7).

The eighteen state farmers' associations are full members of the DBV, together with the Bund der Deutschen Landjugend (Association of German Young Farmers), the German Raiffeisen Association, and the Bundesverband der Landwirtschaftlichen Fachschulabsolventen e.V. (Federal Association of German Technical College Graduates). Individual farmers are not direct members of the DBV, instead they are banded together in the relevant state farmers' associations. The state farmers' associations are themselves subdivided into county and local farmers' associations. Membership of the state farmers' associations is voluntary; however, 90% of the owners of farms over 2 ha are members. In general, membership levels are higher in regions containing medium- and large-sized farms than in regions where the numbers of part-time farmers are above average (Heinze, 1992, p. 63). Currently, the state farmers' associations have a total of 550,000 members (Deutscher Bauernverband, 1998, p. 7).

The DBV's main activity lies in the area of economic policy. The association works in close cooperation with the federal and state Agricultural Ministries, state authorities, and political parties. On the European Union level, the association works through the Comité des Organisations Professionelle Agricoles de la CE (COPA), of which it has been a member since 1958 (Henrichsmeyer and Witzke, 1994, p. 449). Its main external, private enterprise function consists of undertaking measures to improve the image of the agricultural sector. Amongst the association's important internal activities is the provision of services through county branch offices, e.g. counselling, bookkeeping, training. A further activity is the explanation of state agricultural policy to association members.

Institutional Characteristics of German Agriculture

The DBV has competition as the representative of farming interests in the old federal states (formerly West Germany) in the form of smaller associations, such as the Arbeitsgemeinschaft Bäuerliche Landwirtschaft – Bauernblatt e.V. (Working Group Rural Agriculture – Farmers' Paper). Some of these associations, often only regional, joined together in 1988 to form the Dachverband der Deutschen Agraropposition (Congress of German Agrarian Opposition). These interest groups, however, have been of very little importance to date in comparison to the DBV. In contrast to developments in the old federal states, the professional associations in the new states (formerly East Germany) were themselves organised to counterbalance the DBV. The strength and influence of these vary from state to state. The DBV is, however, also attempting to establish a general, united association to represent the farming industry's interests (Henrichsmeyer and Witzke, 1994, p. 13).

Institutions for Information, Counselling, and Administration

Education and Vocational Training

The origins of the acquisition and provision of information by institutions in the field of agriculture go back to the founding of agricultural academies and universities at the beginning of the 19th century. Later these were supplemented by the agricultural schools, which provided vocational training for farmers. Today, the educational facilities in the field of agriculture consist of (Henrichsmeyer and Witzke, 1994, p. 84):

- colleges of applied science and universities training executive personnel for administrative and commercial positions in farming and expanding scientific knowledge in agriculture through research activities,
- agricultural technical colleges providing vocational training for plant and farm managers, preparing them for voluntary mastercraftsman examinations
- agricultural vocational schools providing instruction parallel to practical apprenticeship training, a compulsory component of an agricultural apprenticeship

Existing numbers of state institutions of education and vocational training are as follows (AID, 1997):
- universities with agricultural faculties: 11
- colleges of applied science with agricultural departments: 8
- agricultural technical colleges: 209
- agricultural vocational schools: 580

Administration

Agricultural administration comes under the responsibility of the individual states, which have each erected the required administrative structures to facilitate implementation of state, federal, and common European agricultural policies. In general, the lowest level is the agricultural authorities, which perform sovereign administrative tasks. Here, farmers submit their applications and from here transfer payments are organised. Some states have created other administrative structures by outsourcing tasks to separate, independent organs within the agricultural sector. In this way the professional, self-governing institutions known as the chambers of agriculture were created. The agricultural authorities have been allocated a host of administrative and, increasingly, supervisory tasks. In many cases, these authorities are equipped with agricultural schools and agricultural counselling centres.

As a result of the decreasing number of farms, caused by structural change and budget restrictions, the current policy is to considerably reduce administrative workloads and in conjunction with this, initiate functional change within the individual authorities. Income maintenance policy in agriculture has shifted in recent years from price pegging to direct payment for individual farms, in some cases for individual fields, leading to a greatly increased administrative workload. The execution of the different aid programmes requires considerable administrative efforts. Additionally, there is increasing conflict between individual counselling and the need for supervision, both from the point of view of agricultural policy, and because it is stipulated under common European policy.

The tasks of the chambers of agriculture, amongst others, are to provide counselling for farmers, to support them in the transfer of technical improvements, to help them in rationalisation endeavours, to take responsibility for the education and vocational training of current and future farmers and managers, and to supply economic advice, as well as to give expert opinions on matters relating to legislation, administration, and court decisions (Pacyna, 1998, p. 97). In order to fulfil these tasks, the chambers of agriculture were granted the right to levy fees. Additionally, state support means the chambers also receive public funding. The chambers of agriculture have a long tradition, having been founded as self-governing bodies about two hundred years ago (Verband der Landwirtschaftskammern, 1979, p. 3). The chambers of agriculture have a quasi-autonomous nature. On the one hand, their democratic structure and membership of farming enterprises means they function largely as a self-governing body. On the other hand, they and their activities are controlled by the state, as a result of receiving public funding.

Counselling

Originally, education, information, and counselling were regarded collectively as the central instrument of agricultural policy; the state organised, financed, and provided schools and counselling services. In recent years this strict allocation of responsibility to state organs has been relaxed, so that currently four different, but somewhat parallel, counselling opportunities are available to farming enterprises. In order of lessening state influence, these are the counselling provided by the state, by the rings and working groups, by the private sector, and by other institutions.

Official counselling covers all forms of agricultural counselling organised and largely financed by the state. In many cases, farmers and their families are offered scholastic training in conjunction with the counselling. In recent years, official counselling has been increasingly restricted. Budget considerations, conflicts between counselling and the need for supervision, and especially insufficient counselling competence have been pinpointed as reasons for reducing official counselling.

Next is the increasingly important ring counselling. Farmers band together voluntarily to form associations to jointly employ consultants. Although they vary from one state to another, generally rings exist everywhere and are subsidised by the state. Compared with official counselling, this organisational structure has the significant advantage of eliminating conflicts of interest between consultant and supervisor. The dues-paying members are only interested in receiving efficient counselling, and see the consultant not as a representative of the state, but as one of their own. The average membership of a ring varies from 80 to 150 farming enterprises. There is a growing trend in counselling groups and farm manager working groups towards specialisation in certain branches of production.

Counselling in the Federal Republic of Germany (FRG) is organised differently from state to state. The fundamental principle, however, is that a type of state-funded counselling is available everywhere, but that it contains an increasing number of private sector elements and retains its official status only through joint financing and an institutional connection to the chambers or authorities.

The third option is that of private sector counselling. When considering the total number of farming enterprises, the full-time consultant still plays a relatively unimportant role. As a result of their comparatively high cost, private consultants are only called in under exceptional circumstances, e.g. during preparations leading up to mergers and alliances or counselling very large businesses. It is notable, however, that the demand for private counselling is gradually on the increase; especially in the new states, local structures have led to a dominance by private consultants, although they, too, are often subsidised by the state.

Finally, many other institutions offer farmers and their families counselling services. Along with firms producing and trading in agricultural products and firms processing and marketing them, associations, banks, and churches also offer counselling services.

In summary, Table 2 provides an overview of the structures underlying the counselling services in the FRG. The most important providers of counselling are listed together with their areas of counselling. Since official counselling is provided by the states, regional differences can occur, whilst the services provided by the associations and the private sector remain constant and comparable.

Currently, it can be observed that increased commercialisation and, in part, increased privatisation in the field of agricultural counselling have improved the chance for increased efficiency. To date official counselling by the state and the chambers of agriculture has been provided free of charge, but this field is also experiencing a growing trend towards fees and efficiency.

Table 2: Activities and sponsorship of agricultural counselling in Germany

	State	Chamber	Ring	Private Management Consultant	Farmers' Association	Accounting Agency	Producer Rings	Up- and Downstream Enterprises	Private Specialist Consultant
Business Administration	++	++	++	++		+	+		
Financing	++	++	+	++	+	+			
Applications	++	++	+	++	+				
Marketing	++	++	+	++			++	++	
Law					++	+			+
Taxes						++			+
Economics	++	++	+	+	++				+
Insurance	+	+		+	+				+
Production Engineering							+		
Plants / Animals	++	++	++	++			++	+	+
Agricultural Engineering	+	+	+	++			+	+	
Construction Queries	++	++	+	+					+
Environmental Protection	++	++	+	+	+		+		+
Domestic Economics	++	++			+				

Source: Köhne, 1995, p. 179. ++ = very important + = somewhat important

Institutions Promoting Cooperation in Production

With partial state support, a multitude of cooperatives, known as 'rings', have been formed in German agriculture with the aim of allowing individual farming enterprises to benefit from economies of scale and take advantage of comparative advantages brought by shared experience. These cooperatives consist of machinery and farming rings, producer cooperatives and producer rings, and supervisory rings.

Machinery and Farming Rings

Machinery and farming rings consist of agricultural enterprises which have banded together voluntarily. By sharing their machines, farmers aim to improve capacity utilisation of the machinery and reduce fixed costs.

The rings are structured so that machinery is either purchased jointly, remains property of the ring, and is offered to farmers for use, or the machinery ring acts as an intermediary between farmers, thus bringing together the free capacities of one farmer with the needs of another. The cost rates for the farmer using the machinery are fixed uniformly across the machinery ring.

The machinery ring is organised along the lines of an association, with memberships and a board of directors. Generally, the daily business of the machinery ring is run by a full-time chief executive. The number of members in one machinery ring varies between 100 and 2,400, membership of the Bundesverband der deutschen Maschinenringe e.V. (Federal Association for German Machinery Rings) with approximately 300 considered as an efficient size (Golter et al., 1992, p. 431).

In 1998 there were 292 machinery rings with 200,400 members in the FRG; these rings cultivate 52% of the total arable land in Germany. The machinery rings are mainly located in the south (43% of all machinery rings with 63% of all members are in Bavaria and Baden-Württemberg), to a certain extent as a result of structural characteristics. The advantages of machinery rings are undeniable and are reflected in their growing importance. Whilst the state subsidises some of the organisational costs, state sponsorship of machinery rings is being reduced. Today the rings are becoming increasingly successful even without state support according to the Bundesverband der deutschen Maschinenringe e.V. (Federal Association for German Machinery Rings).

Producer Cooperatives and Producer Rings

Producer cooperatives and producer rings have developed out of the necessity to improve the market position of agricultural producers. According to the Market Structure Law of 1969, producer cooperatives receive state subsidies in the form of start-up help and investment aid (Mühlbauer, 1992, p. 180) and exceptions are made to competition laws to further support them (Mark, 1997, p. 31).

Table 3:
Number of producer cooperatives by type of good, 1998

Type of Good	Producer Cooperative	Type of Good	Producer Cooperative
Cattle for Slaughter and Breeding	47	Potatoes	88
Cattle for Slaughter and Pigs	166	Flowers, Decorative Plants, and Nursery Trees	8
Milk	121	Flowers and Decorative Plants	10
Eggs and Poultry	41	Breeding Cattle	18
Wine	197	Quality Rape	73
Cereals, Oil	28	Others	61
Quality Cereals	344	Total	1,202

Source: *Agrarbericht der Bundesregierung 1999, p. 88*

Crop production takes place according to certain production norms and is carried out in consultation with the other members. The producer cooperatives are often under contract with processing and marketing firms from the next stage of production. The target is to produce and bring to market large, uniform batches of a product. As a rule, producer cooperatives themselves do not possess their own facilities for processing and marketing raw agricultural products. The aim is to jointly provide the goods to processors and/or marketers. In the main, producer cooperatives specialise in one specific branch of production. Currently 1,202 producer cooperatives are listed in the Agricultural Report (Table 3).

Supervisory Rings

The third group of institutions consists of the supervisory rings. These rings organise and carry out specific aspects of production, in general quality and performance control. Farming enterprises join together through the exchange of experiences in order to use uniform approaches to control and improve production.

The formation of supervisory rings in animal husbandry, especially dairy farming and pig farming, is widespread. In the field of dairy farming there are sixteen state supervisory associations. North Rhine-Westphalia and Lower Saxony each have two state supervisory associations (in some federal states the associations are subdivided at the county level) and two milk testing rings can be found in Bavaria and Baden-Württemberg (Arbeitsgemeinschaft Deutscher Rinderzüchter e.V., the working group for German cattle breeders). There are currently approximately 120 supervisory rings active in the field of pig farming. A large portion of the administrative costs are financed by the federal states.

Institutions Promoting Cooperation in Purchasing and Marketing – the Agricultural Cooperatives

Principles and History

For nearly 150 years, a tradition of agricultural cooperatives has existed in Germany. In 1862 Friedrich Wilhelm Raiffeisen founded credit cooperatives as loan bank associations which soon began trading in commodities. (A vivid account of Raiffeisen and his cooperatives can be found in Faust, 1977, p. 323 ff.). They were founded with the aim of utilising economies of scale, enabling the development of market power along the lines of a counterweight and, by working together, to accelerate innovation and implementation of technical advances in processing and marketing. Today these continue to be the purposes of Raiffeissen cooperatives.

Institutional Characteristics of German Agriculture

Cooperatives are based on the principles of self-help, self-government, and responsibility for one's own actions. Their aim is to foster the economic well-being of their members (Art. 1 Sec. 1 Cooperative Association Law). Owners, i.e. investors, and customers should be identical groups of people (known as the principle of identity). As democratically organised enterprises, they remain clearly separate in goals, decision-making processes, and supervision from their competitors on the market. These fundamental principles have remained largely untouched by functional and structural changes in the farming sector and retain their validity today. Nevertheless, the image and character of the cooperatives have changed radically from a self-help organisation for the poor, rural population to modern, powerful institutions.

Structure of Agricultural Cooperatives

After undergoing extensive structural change, 4,221 agricultural cooperatives are grouped together under the umbrella of the Raiffeisen Cooperatives (the following figures are from Agricultural Report of the Federal Government, 1999, p. 27; DRV Online, 1999 unless otherwise noted). Their total turnover was approximately 75 billion DM in 1998. Around half of the total purchasing and sales turnover generated by farmers is through their cooperatives. Thus, the rural cooperatives enjoy a global market share of 50%.

The structure of the agricultural cooperatives can be divided into three distinct levels. At the local level, there are 4,184 cooperatives with three million members. The next level is the regional level, with corresponding regional central enterprises which in turn support national enterprises at the federal level. Associations can be found at both the regional and the federal levels (Figure 1).

Figure 1:
Structure of rural cooperative organisations, 1998*

Federal Level
Nationwide central enterprises
German Raiffeisen Goods Centre Ltd
German Milk Counting House Ltd
German Wine Cooperative e.G.
German Raiffeisen Association e.V. , Bonn
Regional level
33 Regional central enterprises
9 Main Cooperatives
6 Dairy Cooperatives
4 Cattle and Meat Centres
3 Central Wineries
11 Others
Regional Testing Associations
Local level
4,184 Rural Cooperatives with 2,957,000 members, of which:
566 Credit Cooperatives with commodity trade, 1,966,000 members
576 Agricultural Purchasing and Marketing Cooperatives, 160,000 members
455 Dairy Cooperatives, 199,000 members
140 Fruit, Vegetable and Horticultural Cooperatives, 50,000 members
278 Wine Cooperatives, 66,000 members
127 Cattle and Meat Cooperatives, 119,000 members
837 Agricultural Cooperatives, 56,000 members
1,205 Other Cooperatives, 345,000 members

Sources: Aschhoff and Henningsen, 1995, p. 90; Zerche et al., 1998, p. 21; DRV Online, 1999.
* Preliminary membership figures as of May 1999.

Institutional Characteristics of German Agriculture

Selected Cooperative Types in German Agriculture

Credit cooperatives with commodity trade are an exceptional form of cooperative. They can be found principally in southern Germany and still fit the ideal of the complete cooperative as Raiffeisen saw it: the cooperative finances the farmer's procurements and the purchase of livestock and then helps process and market the harvest and animal produce. It has been observed that the economic strength of credit cooperatives that are also involved in commodity trade is waning, and they will likely cease to exist within the next few years. The decline of these cooperatives is a result of several factors. Today farmers can receive loans from various sources. Additionally the small size of these cooperatives leads to high unit costs, meaning they are no longer competitive in both banking and commodity trade.

Of more importance are the agricultural purchasing and marketing cooperatives, which offer a specific product range and achieve a market share of between 40% and 55% in the markets for cereals, fertilisers, pesticides and herbicides, and farm machinery (DG-Bank AG, 1998, p. 32). Significant movement towards concentration has been observed in this sector. At the same time, the cooperatives are undertaking diversification strategies which have led to most of these enterprises becoming increasingly active in the retail industry for gardening, domestic and construction products, and fuels, as well as the wholesale trade in construction materials (GenoLex, 1992, p. 74). The major problem associated with this group is that traditional agricultural business is on the decline and farmers are being forced to agree to an increasing number of investments which are important for rural regions, but are not important for agricultural production.

Cooperatives are commonly involved in the production and processing of milk. Dairy cooperatives have also been affected by the structural changes that have taken place in up- and downstream enterprises. From 1978 to 1998, the number of dairy cooperatives fell by 80%. At the same time, based on the quantity of milk produced, their market share doubled. Although the concentration process has slowed in the 1990s, the number of co-operations and strategic alliances, which in some cases are preliminaries to merging into a cooperative, has grown (DG-Bank, 1998,

p. 34). Each of the four largest businesses process between 1 and 4 million tonnes per year.

Cooperative dairies were very heavily involved in the production of goods eligible for subsidies (butter, dried skimmed milk), but now focus their product lines more directly on market conditions. Rising demand for fresh milk products induced the cooperatives to expand fresh milk production and to extend their product range (Bundesministerium für Ernährung, Landwirtschaft und Forsten, 1999, p. 22). Developments on world markets are having a stronger influence on the future prospects for the dairy cooperatives due to the fact that companies' exports are increasingly important (DG-Bank, 1998, p. 34; DRV, 1998, p. 34).

Cooperatives in specialised cultivation, i.e. fruit, vegetable, horticultural, and wine cooperatives, have spent the last few years orienting themselves more to market developments and have consequently come to perform the central marketing function for many producers, and enjoy correspondingly high market shares. Nevertheless, they are still the segment of primary cooperatives with the weakest turnover. The level of commitment to these cooperatives is relatively high, with members closely identifying themselves with their respective cooperatives, accepting stipulated production regulations.

The fruit, vegetable, and horticultural producers must accept a stronger marketing concentration as a result of the similar trend amongst competitors and in the grocery trade (DG-Bank, 1998, p. 38). Structural change in the wine cooperatives is being countered with improvements in the organisation of marketing structures at the regional level. The number of German wine producer cooperatives has fallen continuously; only 30% of wine producers belonged to a cooperative in 1997 (DG-Bank, 1998, p. 35).

The cattle and meat cooperatives have been plagued by considerable economic difficulties for many years. The number of cooperatives has fallen over the past decade. The reduction of excess capacity in the slaughterhouses, radically increased intervention holdings of beef, and strong European competition on the cattle and meat markets, together with failures to deliver on time, have led to permanent economic difficulties within this sector. Currently, attempts are being made to counteract these difficulties with capacity adjustments, concentration processes, the increased implementation of supply contracts, and a definite trend towards exporting. Additionally, the introduction of brand name meat programmes has served to solidify the cooperatives' position as an important partner for the grocery and butcher trades (DG-Bank, 1998, p. 35). The cattle and meat cooperatives are also at the forefront in implementing network systems between the production of cattle for slaughter and the grocery retailing sector. Such networks consist of uniform standards for hygiene, disease prevention, and quality (DRV, 1998, p. 38).

Finally, there is a broad range of other service cooperatives for the agricultural sector, which offer member enterprises many and varied services, from chilling facilities to technical assistance. For many years these cooperatives have been involved in the cultivation and marketing of biomass. Furthermore, the cooperatives also perform pilot functions in marketing biodiesel. Particularly in rural regions, the cooperatives have completely taken over the task of providing consumer goods for rural populations.

The farm production cooperatives play a special role. Their orientation is not on the up- or downstream stages, but on agricultural production itself. They can be found in the new federal states and are, as a rule, the successors of the former socialist production cooperatives. These cooperatives are also undergoing change. Whilst after the fall of communism, cooperatives concentrated on commercial production, in recent years they have expanded their activities and are now active in processing and marketing their products and are increasingly diversifying (Grosskopf, 1996, p. 76).

The Future of Agricultural Cooperatives

Future development of agricultural cooperatives will be influenced by numerous factors. One major influence will be the reorientation of European agricultural policy as a result of the Agenda 2000 and the resultant accelerated structural change for farming enterprises, which will impact the membership structures of the cooperatives. Globalisation and market liberalisation, which have up- and downstream effects, will also strongly influence the cooperatives. To remain competitive it will be necessary to focus on European and global markets. The integration of Eastern Europe into the European Union will also bring new challenges and markets for the cooperative organisations. Competitiveness and the ability to establish themselves in and conquer new markets will become decisive factors for their survival. At the same time, technical progress in agriculture, linked with the increased use of biotechnology, will not leave cooperatives untouched. The highly capital-intensive nature of biotechnology will increase contractual cultivation and strengthen interaction with multinationals and global corporations. The result will be a fall in the number of independent family businesses and thus a reduction in membership numbers in the traditional cooperatives (Frankfurter Allgemeinen Zeitung, 1999).

Structural change in agriculture itself has restricted the scope of activity of rural cooperatives. Constraints are brought by investors' demands, the concentration strategy pursued by the cooperative partners, both on the supply side (suppliers of farming machinery, fertilisers, pesticides and herbicides) and on the marketing side (food industry and grocery trade). The necessity for adjustment arising from these underlying conditions is characterised by two problem areas. First, farmers are limited in their willingness to invest the capital necessary for adjustment, if they wish to invest at all; they want 'good' prices without increasing their capital investment. Second, the decision-making processes in the cooperatives are relatively ponderous; democratic processes mean implementation takes a long time. Both problems are currently under discussion and solutions are being sought.

References

AID (Ed.) (1997). Bildungstätten im Agrarbereich der Bundesrepublik Deutschland, Bonn.

Aschhoff, Gunther, Henningsen, Eckart (1995). Das deutsche Genossenschaftswesen: Entwicklung, Struktur, wirtschaftliches Potential, 2., völlig überarbeitete und erweiterte Auflage, Frankfurt (Main).

Bundesministerium für Ernährung, Landwirtschaft und Forsten (Hrsg.) (1999). Struktur der Molkereiwirtschaft. Reihe: Daten-Analysen. Bonn.

Bundesministerium für Ernährung, Landwirtschaft und Forsten (Hrsg.) (1999). Agrarbericht der Bundesregierung 1999. Bonn.

Deutscher Bauernverband (Hrsg.) (1998). Geschäftsbericht des Deutschen Bauernverbandes für das Jahr 1997. Bonn.

DG-Bank AG (Hrsg.) (1998). Die deutschen Genossenschaften 1998. Bericht. Frankfurt (Main).

DRV (Hrsg.) (1998). Jahrbuch 1998, Bonn.

DRV Online (1999): http://www.raiffeisen.de/profil/statistik.htm vom 19.07.1999.

Faust, Helmut (1977). Geschichte der Genossenschaftsbewegung: Ursprung und Aufbruch der Genossenschaftsbewegung in England, Frankreich und Deutschland sowie ihre weitere Entwicklung im deutschen Sprachraum, 3., überarbeitete und erheblich erweiterte Auflage, Frankfurt (Main).

Frankfurter Allgemeinen Zeitung (1999). Konzentration der Unternehmen in der Biotechnologie. July 12, 1999.

Golter, F. in: Mändle, E., Swoboda W. (1992). Genossenschafts-Lexikon; Deutscher Genossenschaftsverlag e.G., Wiesbaden, S. 431.

Grosskopf, Werner (1996). Diversifizierung – eine erfolgreiche Strategie für Agrargenossenschaften? in: Institut für Genossenschaftswesen an der Humboldt-Universität zu Berlin (Hrsg.): Entwicklungschancen ländlicher Genossenschaften in den östlichen Bundesländern. Berlin, S. 74-79.

Heinze, R. G. (1992). Verbandspolitik zwischen Partikularinteressen und Gemeinwohl: Der Deutsche Bauernverband. Gütersloh, S. 63.

Henrichsmeyer, Wilhelm and Witzke, Heinz Peter (1994). Agrarpolitik, Band 2: Bewertung und Willensbildung, Stuttgart.

Köhne, Manfred (1995). Struktur und Weiterentwicklung des landwirtschaftlichen Beratungswesens in Deutschland, in Ausbildung & Beratung 11/95, S. 179.

Maisack, S. (1995). Leitbilder und Mitgliederverhalten in Verbänden. Frankfurt a. M..

MÄNDLE, E. (1983). Willensbildung in der Agrarpolitik. In: Fredeburger Hefte Nr. 9, (Ed.) Deutsche Landjugend-Akademie Fredeburg, Fredeburg, S. 73ff.

MÄNDLE, E., SWOBODA, W. (Hrsg.) (1992). GenoLex Genossenschafts-Lexikon, Wiesbaden.

Mark, Steffen (1997). Erzeugergemeinschaften in Baden-Württemberg, in: Landinfo 6/97, S.31.

Mühlbauer, F. in Mändle, E., Swoboda W. (1992). Genossenschafts-Lexikon; Deutscher Genossenschaftsverlag e.G., Wiesbaden, S. 180.

Pacyna, Hasso (1998). Agrilexikon, Hannover, S. 97f..

Schnieders, R. (1998). Von der Ablieferungspflicht zur "Agenda 2000". In: Deutsche Bauernkorrespondenz, Heft 3, S. 4 - 10.

Verband der Landwirtschaftskammern (Hrsg.) (1979). Die Landwirtschaftskammern in der Bundesrepublik Deutschland, Bonn, S. 3f..

Zerche, Jürgen; Schmale, Ingrid; Blome-Drees, Johannes (1998). Einführung in die Genossenschaftslehre: Genossenschaftstheorie und Genossenschaftsmanagement, München.

Chapter 9

The Hierarchy of Agricultural Policies: European Union, Federal Republic, and Federal States

Carsten Thoroe
Institute for Economics, Federal Research Centre for Forestry and Forest Products Hamburg

Introduction

It is difficult to describe and analyse the distribution of responsibilities in relation to agricultural policy in Germany. In this area, the institutional responsibilities of the various governmental levels (EU (or EC, respectively), federal government, and federal states) are mixed, with different functional spheres of responsibility such as decision making, implementation, and finance. With the founding of the European Economic Community (EEC) and the decision to have a common agricultural policy, institutional responsibilities were transferred to the Community level. In the course of the development of the common system of market regulations, the operation of the Common Agricultural Policy (CAP), and progress in the process of integration, these have been supplemented in content. However, the Community responsibilities in these areas are primarily decision making and financing. Responsibility for implementation has remained at the national level.

The political interlinking of the EU and the national levels is complicated by an additional governmental level in Germany, a consequence of the federal structure. Frequently, the EU and both the federal government and federal states are included in the decision-making process and the implementation of measures related to agricultural policy. Increasing bureaucratisation and broadening policies have led to substantial resistance, especially in the federal states where clamorous calls are made for more state-level responsibility.

This chapter will describe the interlinking of the different levels and the mixing of responsibilities, as well as the resulting problems and obstacles to agricultural policy reforms. First, the federal structure of the Federal Republic of Germany (FRG) and its development will be outlined; the interlinking of the different levels and the mixing of responsibilities creating a cooperative federalism, is quite marked in Germany. (Strauss, 1969). Following this, an attempt will be made to outline the distribution of responsibilities under European Community (EC) agricultural policy. Then the question will be raised as to the extent to which a modification of the distribution of responsibilities is required as a result of reform in agricultural policy, increased integration of agriculture in international world trade, as well as of the anticipated eastward expansion of the EU. Finally, some challenges for agricultural policy will be inferred from these considerations. The concluding remarks are based, in particular, on the recently published opinion of the Scientific Advisory Board of the Federal Ministry for Food, Agriculture and Forestry concerning the 'allocation of responsibility for agricultural policy in the EU' (WB-BML, 1998).

The Federal Structure of the Federal Republic of Germany

The public finance system, the obligation for public responsibilities, and the levying of taxes is subdivided into three levels: federal government, federal states, and local government (including the associations of municipalities). The allocation of responsibility is subject to the principle of subsidiarity, i.e. the implementation of responsibilities should be, as far as possible, decentralised. The FRG constitution grants a strong position to the federal states in relation to the federal government. The fulfillment of typical governmental responsibilities is basically a matter for the federal states; restrictions arise only where the Grundgesetz (GG, or Constitutional Law) makes a different provision (Art. 30 GG). Likewise, the federal states have legislative rights wherever the GG does not expressly grant such power to the federal government (Art. 70 GG).

The areas in which the federal government has responsibility for legislation are :
- relationships with other states
- the monetary system including the minting of coins and currency
- the post and telecommunication systems
- protection of industrial property rights and copyright and publishing rights

In some areas there is concurrent legislation at the state level, the federal government only stepping in when it is in the interest of the state as a whole to have general regulations. This concurrent Legislation at the state level is limited to 24 specially nummerated areas (Art. 74 GG). One of these areas is 'the support of agricultural and forestry production, safeguarding food, the import and export of agricultural and forestry products, deep-sea fishing and inshore fishing, and coastal protection' (Art. 74, Nr. 17 GG).

In fact, the decentralised lines of the constitutional law have lost their contours over time. With the aid of concurrent legislation, the federal government has extended its range of responsibilities, so that the responsibility for making decisions has become more centralised. In addition, in a few areas the federal government has increased its influence by participating in the financing of measures for which the federal states were responsible. As a result of the 1969 reform of the public finance system, the federal government's share in responsibility for agricultural structure and coastal protection, regional support, and construction of universities is actually rising. This was made according to the constitution through the introduction of 'joint tasks' (Art. 91a GG). The intention was to mobilise 'the financial resources and authority of the state as a whole by means of a coordinated planning and financing system between the federal government and the federal states to accomplish these 'joint tasks' (Scharpf, 1994, p. 16).

In the case of the joint task 'improvement of the agricultural structure and coastal protection' (GAK), the federal government and the federal states act together in the planning and financing of aid measures. A planning committee, consisting of sixteen representatives of the federal government and sixteen of the federal state governments, votes each year on a skeleton plan, in which the rules governing the prerequisites, type, and intensity of the aid measures are laid down, as well as the distribution of the resources available to the recipients of the measures and to the federal states. The responsibility for implementing these measures is placed exclusively with the federal states, which issue their own guidelines for this purpose.

Table 1:
Tax Revenues by Type and Recipient, 1998

	Billion DM	% of total
Type of tax		
Shared taxes[a]	600.7	72.1
Federal taxes	130.5	15.7
Customs duties	6.5	0.8
Taxes levied by the federal states	37.3	4.5
Municipal taxes	58.0	7.0
Total	833.0	100.0
Recipients of the taxes, according to federal supplementary appropriations		
Federal government	341.5	41.0
Federal states[b]	344.1	41.3
Municipal authorities	105.1	12.6
European Union	42.3	5.1
Total	833.0	100.0

[a] Including trade tax and increased 'Gewerbesteuerumlage' (participation of federal and state governments in the municipal trade tax, Art. 106 VI GG). [b] Including revenue from municipal tax of the city-state federal states (e.g., Hamburg, Bremen)353.9 billion DM (42.25%).
Source: BMF, 1999a; 1999b

With the 1969 public finance reform, the principle of fiscal equivalence was emphasised, since according to Art. 104a GG the federal government and the federal states separately bear the expenses which result from the implementation of their tasks. At the same time, however, exceptions were codified; for the joint tasks just mentioned, for laws governing cash benefits (Art. 104a, Para. 3 GG), and for financial aid for particularly important investments (Art. 104a, Para. 4 GG), the financial participation of the federal government is laid down. Thus, financial responsibility originally conceived as highly decentralised has also acquired more centralised features. While the different governmental levels were initially each allocated their own revenues, now joint or shared taxes predominate. Several levels of government share in the revenues from almost all large taxes. In 1998, more than 72% of tax revenue in Germany went to such shared taxes; the proportion of purely federal taxes constituted 16%, and those of purely federal state taxes constituted only 4.5% (Table 1).

Taking the developments on the revenue side and the expenditure side together, in the national sphere in Germany there is the impression that the various governmental levels form a sturdy 'network'. Only at the level of implementation are there quite pronounced decentralised elements. There has been much critical discussion since the late 1970s of this cooperative federalism, the result of the networking of government levels (Scharpf, 1977; Lehner, 1979; Fürst et al., 1984). However, in principle it has not been changed.

Distribution of Responsibility in Relation to Agricultural Policy in the Course of European Integration

In the years following the Second World War, agricultural policy in the FRG was, as elsewhere in Europe, initially strongly oriented to safeguarding the food supply (Henrichsmeyer and Witzke, 1994, p. 542 ff). To this end, a system of market regulations was established for important agricultural products, founded partly on institutions created before the war. In contrast to the general basic decision in favor of a social market economy, the implementation of FRG agricultural policy was orientated on political intervention (Niehaus, 1957; see also Koester, this volume).

The highly regulated sectors of the economy (coal, steel, agriculture, transport, nuclear energy) had great importance in the European integration negotiations. While for coal and steel, as well as nuclear energy, independent contracts were concluded for founding European Community organisations (European Community for Coal and Steel, European Nuclear Power Community), transport and agriculture were areas for which a common policy should be developed within the framework of EEC (EEC contract 1957). At the same time, the foundation of a joint organisation of agricultural markets was established, in order to achieve the goals for a common agricultural policy, as formulated in the EEC contract.

Agricultural Markets Policy

In the initial phase of the EEC, the work of the Community was concentrated, above all, on the joint agricultural policy, primarily with the development of policies in relation to the market and prices. Of the three possible forms of organisation provided for in the EEC contract (common regulations on competition, binding coordination of the various market regulations of the individual states, European regulation of markets), the one that was realised was that which allocated the greatest growth in decision-making responsibility and financial responsibility to EU organs: European market regulations. EU market regulations have been developed for all important agricultural products, replacing national market regulations valid up until then in the individual Member States. In 1962, the first EU market regulation was put into force, governing the cereals market. Step by step in the following years, common market regulations were agreed upon for most agricultural products, variously formulated according to the type of product and specific market conditions. The fundamental principles of a common agricultural policy can be defined as follows:

- unity of the common market
- preference for production within the market, as opposed to importing (Community preference)
- financial solidarity

In order to finance the CAP, a European Alignment and Guaranty Fund for Agriculture (EAGFA) was set up, which would fund expenses resulting from the EC agricultural policy. Even today this fund still dominates the EU budget (Table 2). Agricultural policy was viewed as the trailblazer for European integration, indeed it was the only area in which the allocation of responsibilities was so far-reaching. This was the outcome of the EEC contract that, despite very different national problems and interests,

Table 2:
Separable Returns and Contribution Payments by EU Member States, 1997*

	Returns				Contribution Payments		Share of community GNP	Participation coefficients	
	In total		Including: Agricultural policy					Agr. Policy (4):(7)	Contribution payments (6):(7)
	Million ECU	%	Million ECU	%	Million ECU	%			
	(1)	(2)	(3)	(4)	(5)	(6)	(7)	(8)	(9)
Luxemburg	118	0.2	23	0.0	149	0.2	0.2	–	1.00
Belgium	1,83	2.6	983	2.4	1,923	3.1	3.1	0.77	1.00
Denmark	1,54	2.2	1,236	3.0	1,217	2.0	1.9	1.58	1.00
Austria	1,37	1.9	861	2.1	1,856	3.0	2.6	0.81	1.15
Germany	10,141	14.2	5,778	14.2	17,785	29.1	26.0	0.55	1.12
Netherlands	2,52	3.5	1,757	4.3	3,109	5.1	4.5	0.96	1.13
France	12,214	17.1	9,149	22.5	11,635	19.0	17.2	1.31	1.10
Italy	8,51	11.9	5,091	12.5	7,547	12.3	14.2	0.88	0.87

	Returns				Contribution Payments		Share of community GNP	Participation coefficients	
	In total		Including: Agricultural policy					Agr. Policy (4):(7)	Contribution payments (6):(7)
	Million ECU	%	Million ECU	%	Million ECU	%			
	(1)	(2)	(3)	(4)	(5)	(6)	(7)	(8)	(9)
Great Britain	7,04	9.8	4,400	10.8	5,884	9.6	16.1	0.67	0.60
Finland	1,10	1.5	571	1.4	918	1.5	1.4	1.00	1.07
Sweden	1,18	1.7	747	1.8	1,963	3.2	2.7	0.67	1.19
Ireland	3,35	4.7	2,034	5.0	462	0.8	0.8	6.25	1.00
Spain	11,279	15.8	4,606	11.3	4,736	7.7	6.6	1.71	1.17
Portugal	3,78	5.3	657	1.6	923	1.5	1.2	1.33	1.25
Greece	5,53	7.7	2,731	6.7	1,015	1.7	1.5	4.47	1.13
Total	71,546	100.0	40,623	100.0	61,121	100.0	100.0		

Note: * Figures rounded and do not include administrative costs.
Source: WB–BMWi, 1999; own calculations

developed a genuine 'Europeanised' political domain. While in other areas of politics insufficient readiness to surrender sovereignty put a brake on European integration, an attempt was made in agricultural policy to take into account opposing national interests in the joint political decisions. A 'safety valve' was achieved with the introduction of Monetary Compensation Amounts although this breached the basic principle of the common market. Thus governments could react to developments in national economies and conflicts that arose as a result of a common agricultural policy. Otherwise, divergent national interests would be compensated chiefly by compromises made during the negotiations, to which finally all Member States could agree. As a result, agricultural interests in the Community inevitably dominated these compromises, frequently at the expense of consumers and taxpayers as well as of third world countries.

Agricultural Structure Policy

After the EC extended its responsibility in the area of agricultural market policy, transferring responsibility for decision making and finance from the federal government to the EC level, in the late 1960s the EC Commission began to turn to problems and policy measures related to agricultural structure. In 1968, it presented a 'Memorandum Concerning the Reform of Agriculture in the European Community' which has become known, after the Commissioner for Agriculture at the time, as the 'Mansholt plan'. The plan encountered strong criticism, above all from professional representation groups, and also received little support in the political decision-making committees. Nevertheless, it laid the groundwork for the first steps towards a common agricultural structure policy. In 1972, an independent EC agricultural structure policy, to provide support for the CAP, was set up and had three main emphases:

- selective investment assistance for individual enterprises 'capable of being supported' (Directive 72/159)
- annuities and premiums for the surrender of land to promote the early retirement of those employed in agriculture with simultaneous structural improvement of the land thereby released (Directive 72/160)
- socioeconomic advice and retraining subsidies for promoting a change of occupation for farmers (Directive 72/161)

These priorities were similar in direction to the German national agricultural structure programme which sought improvement in the competitiveness of agriculture as a result of accelerated structural change. In Germany the implementation of decision-making and financial responsibilities of the EC in agricultural structure policy had effect not primarily at the level of federal government, since this area had originally been the obligation of the federal states. The federal government was first assigned a constitutionally regulated participatory responsibility in the common tasks as a result of the 1969 public finance reform. Thus the already quite complicated interlinking of responsibilities for joint tasks between the federal government and the federal states gained an additional level as a result of the transfer of some responsibilities to the EC. The EC guidelines lay down the framework which is to be put into concrete form and implemented at the national level. At the national level, the federal government and the federal states are jointly required to carry out the joint tasks. Implementation is ultimately a matter for the federal states, which partly include subordinate authorities, such as agricultural boards. In turn, the EC, federal government, and federal states participate in the agreed upon financial distribution formula. The area of education and consultation is not the domain of the federal government. Here, direct responsibility of the federal states was transferred to the EC level, without participation and codetermination rights being granted to the federal states.

Parallel to the support for changes in agricultural structure, in the mid-1970s, measures for supporting agriculture in disadvantaged areas were agreed upon (Directive 72/268). These measures take up at the EU level something that was already the object of national support in Germany, although with a somewhat different orientation.

At the end of the 1980s, new emphasis was placed on fundamental reform of the structural funds. In this reform, funds for EC structural policy were substantially increased in order to reduce the differences between individual regions, with particular attention to least-favoured regions. In addition, the utilisation of resources in the various funds (European Fund for Regional Development, EFRD; European Social Fund, ESF; EAGFA - Department for Alignment; Financial Instrument for the Alignment of Fishing, FIAF) was to be better coordinated- and harmonised with the national activities of the individual Member States (Henrichsmeyer and

Witzke, 1994, p. 556). This means that the intensive interlinking between the EC, the federal government, and the federal states in agricultural structure policy increased substantially.

For the EU structural fund, seven targets were given priority: four so-called geographical targets and three so-called horizontal goals. Responsibility for implementation of the programme lay with the federal states. The EC Commission was, however, represented in the committees that also evaluated the programme. Only coordination, communication, and evaluative functions were given to the federal government. While the federal states implemented parts of their aid programmes via the joint tasks GAK and 'improving the regional agricultural structure', the federal government participated in financing, providing 60% and 50% of the expenses, respectively (WB-BML, 1998, p. 17).

Following the Agenda 2000 reform, structural aid was streamlined; instead of seven categories of objectives, there will be only three from the year 2000 on. The agricultural structure policy will be integrated into a comprehensively organised policy of supporting the development of rural areas. The basis for this support is the development plans for rural areas prepared at the regional level (in Germany, by the federal states) and presented to the EU Commission by the Member States. Consequently, responsibility at the national level for decisions was strengthened. This also applies to the possible reduction of direct payments by up to 20% in the market area according to specific criteria (manpower resources, standard cover sum, total premiums collected) and the utilisation of community funds thus saved for measures to support rural areas. In addition, national limits were laid down for direct payments in the case of milk and beef. With the streamlining of the structural policy and the incorporation of agricultural structure policy into the integrated support of rural areas, a course has been set to make rural development, in conjunction with the expansion of the agricultural environmental programme, the second pillar of the CAP.

Agricultural Environment Policy

The uniform European act that came into force in 1987 extended the regulatory authority of European institutions to some areas in which the EC was already somewhat active, particularly in cases where this authority had been based on the general authorisation clause of Art. 235 of the EC contract. In relation to agricultural policy, it is important to stress that environmental policy was incorporated into the catalogue of areas in which the Community was active (Art. 3 (i) EC contract). In recent years, the EC has set up frameworks in various ecopolitical areas which are binding for the Member States. It has issued some quite concrete regulations and norms regarding the environment, which are to be realised directly in national law and are of substantial importance for agriculture (e.g. drinking water guidelines). In the area of ecopolitics the EC has in general authority in decision-making only. The Member States take care of implementation and financing; in Germany, this is largely dealt with at the federal state level.

In the area of agricultural ecopolitics, the situation is a little different. In this case, the EC participates financially in environmental programmes in accordance with (EEC) Directive 2078/92. The federal states have quickly taken advantage of this of support, briskly bringing existing federal state programmes into the EU programme and launching new programmes (Plankl, 1996). With these programmes, the mixing of responsibilities across the EC, federal government, and federal state levels is particularly clear:

> The federal states plan programmes which are specific to particular areas or which are comprehensive programmes extending over several years, the Federal Government refers these plans to the EC Commission, which evaluates these and decides on the question of approval.
>
> The programmes which have been approved will be financed by the federal states and the EU in the ratio of 50:50; in the so-called Target 1 areas in the ratio of 25:75.
>
> The programmes are implemented by the federal states. Reporting to the EU is carried out via the Federal Government. In so far as the measures are implemented within the framework of the GAK, the Federal Government is included both in the planning, as well as also in financing. The substantial mixing of levels with reference to planning and decision making, implementation, financing and control results in a distortion of the incentive structures and poses substantial problems regarding efficiency. (WB-BML, 1998, p. 34)

The Hierarchy of Agricultural Policies

The Agenda 2000 decisions have certainly strengthened the authority to make decisions at the national/regional level but, in principle, little has changed with regard to the interlinking of responsibility across the EU, the federal government, and the federal states.

Agricultural Social Policy

The area of agricultural social policy has remained a national responsibility. Here, the federal government has made use of its legislative power and developed a differentiated system of independent agricultural social security, which is not integrated into the general statutory social insurance system (Henrichsmeyer and Witzke, 1994, p. 560). This system of agricultural social security is largely supported by subsidies from the federal budget. In 1997, federal government expenditures for agricultural policy totalled approximately 10.7 billion DM, about 7.7 billion DM was allotted to agricultural social security.

Redistributive Aspects of the EAGFA

Overall, a wide range of agricultural policy responsibilities has been transferred from the national level to the EU, particularly in the area of agricultural market policy. Also, agricultural market policy was strongly shaped by a price policy that was aligned to agricultural income levels (Koester, this volume). To support this policy, at the beginning of the 1970s an independent EC agricultural structure policy was introduced. As a result, in Germany not only the responsibilities of the federal government, but also those of the federal states were transferred to the EU. Up until 1992 the federal government could transfer not only its own jurisdiction but also that of the federal states to the European level, without the agreement of the federal states or the Bundesrat (or 'Federal Council', the upper house of parliament representing the federal states in the legislative process) (Art. 24, Para 1 GG). For the first time since 1992, the revised Article 23 GG provides for the participation of the federal states via the Bundesrat, when their sovereignty rights are affected by the surrender of responsibilities (Wiedmann, 1996, p. 352). In the course of this, the EC has increasingly extended its influence on agricultural policy.

However, agricultural social policy and agricultural tax policy have remained essentially within the domain of Member States.

With the extension of its responsibilities, the EC has pressed forward into areas which, to some extent, were already covered by national measures or programmes. The prospect of financial support from the EAGFA may have encouraged individual Member States to agree to the annual negotiation pacts and also with the extension of EC responsibilities. EC agricultural policy makers have been able to open up an area which was unwaveringly protectionist in its orientation despite some vehement criticism. With small concessions here and there, even fundamental conflicts in EC agricultural policy could be skated over, with the objectives of stability and growth, and the furtherance of EC goals in financial, trade, foreign or development policy. The rapidly increasing costs have been accepted as well, although reluctantly (SRW, 1980, p. 176 ff). In anticipation of political integration, EC agriculture policy was from the very beginning largely transferred to the European level. The far-reaching political integration objectives of the 1950s and 1960s have not been realised. The financial limit defined for the EC as a whole was rapidly reached by the substantial expansion in expenses for the CAP. In the 1970s, these increased more than sixfold; in 1979, more than three-quarters of all EC expenditures were allocated to the EAGFA. Not before the covering of the increase in expenses, an institutionalised requirement by the 1984 resolution regarding budget discipline and the 1988 agricultural budget guideline, did the expansion of expenses for agriculture come under control. It was possible to reduce the share of expenses for agriculture somewhat, but they remained dominant in the EU-budget.

Despite the abundant responsibilities the EU level has assumed, its importance in relation to public finance has remained small (Diagram 1). This stands in contrast to expenditure for agricultural policy where the EU level dominates (Diagram 2). Because of the volume of resources, financing has became very important, both in EC agricultural policy discussions and in the EC negotiations. In overall EC agricultural policy, and in terms of individual measures, it can be assumed that the extent to which the individual Member States participate and contribute has a substantial influence on the negotiations. Of the separable returns from

Diagram 1:
Expenditures by the various levels of government, in billion DM, 1997

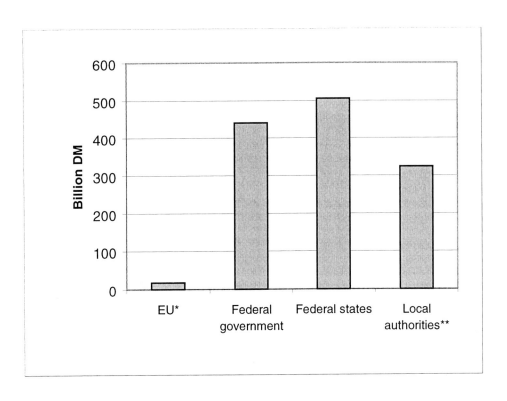

* Payments from the total EU budget to Germany.
** Local authorities, local authority associations.
Source: SRW (1998), p. 372 ff; Deutsche Bundesbank, 1999, p. 61

the EC budget to the individual Member States, 40.6 billion ECU (57%) were allotted to agricultural policy (Table 2). The individual Member States shared in these returns in quite different ways. This participation model is the result of many years of intensive negotiations, in which there were attempts to incorporate divergent national interests. The required majority was sought after in protracted negotiations in the EC Council of Agricultural Ministers, and was found in the form of packet solutions, often based on the smallest common denominator. Similarly, in the attempts to reform EC agricultural policy up until the end of the 1980s, little can be attributed to the power of persuasion of scientific arguments or insights of the Council of Agricultural Ministers; reforms arose, above all, from the pressure from restrictions brought by expenditure limits.

Diagram 2:
Expenditure by the various levels of governement for agricultural policy in Germany in billion DM, 1996*

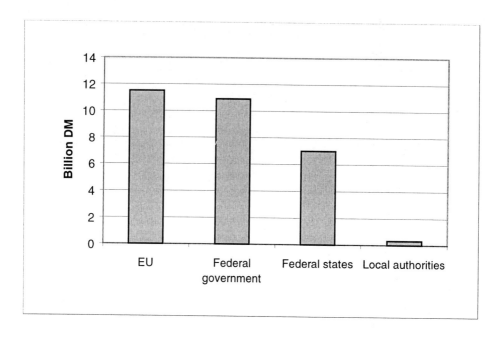

* Expenditure for food, agriculture, and forestry policy, including federal government expenditure for agricultural social security (7.7 billion DM).
Source: Agrarbericht, 1998, p. 92; Statistisches Bundesamt, 1999, p. 492

The southern Member States of Greece, Spain, and Portugal, as well as Ireland, Denmark, and France have especially benefitted from the EC budget via the agriculture policy. This is clearly shown when one views the proportion of returns to the Member States in relation to the proportion of the common social product. This quotient is referred to in column 8 of Table 2 as the participation coefficient for agricultural policy. Germany, Great Britain, and Sweden had the least share in these returns. With regard to contributions of Member States to the EC budget, the participation model is substantially more uniform than the expenditure side (Table 2, column 9).

It was already apparent in the 1960s that marked differences in national problems and interests would hinder further steps towards integration. These differences have increased substantially with the growth in EC Member States from 6 to 9, then to 12, and now at 15. The philosophy regarding integration, as well as agricultural policy objectives have undergone a change. In addition, the eastward expansion of the EU is on the agenda, and will again substantially broaden the spectrum of problems and interests. Therefore, in conjunction with a reform of decision-making structures both in the EU, and in Germany, it appears the time is right to reflect on a change in the distribution of agricultural policy responsibilities, in order to come out of the 'state of political entanglement'. Scharpf (1994) explained this 'state' thus:

> The institutionalised participation of governmental units in the shaping of intentions has inhibiting effects on the power of the superordinated decision-making levels, which can, however, in view of the crisis-ridden and changeable situation with regard to problems, markedly limit the ability to solve problems in politics. They can therefore be described as 'pitfalls', because within the so-described structure, the protagonists can, indeed, possibly repeatedly negotiate more or less satisfactory compromises in relation to the matter in question, but appear to be hardly in a position to produce changes in structure, which could also guarantee, even with a lack of consensus, a high degree of ability to act.

Approaches to a Revision of the Distribution of Responsibilities in Agricultural Policy

Starting from the principle of subsidiarity expressly laid out in the Maastricht Treaty, the orientation to the economic theory of federalism can provide a direction for attempts at a new distribution of responsibilities between the different levels of government (WB-BML, 1998).

In market and income policies, which up until now have been closely interwoven, the EC responsibilities are clearly oriented to safeguarding the operation of the common market. WTO negotiations, agricultural market organisation, and market and price policies are crucial to the successful operation of the common market and are, according to the criteria which have been discussed, clearly within the sphere of EU responsibility. Possibly, closely defined manoeuvring room in the implementation of specific measures could be allowed to Member States. In principle, this also applies for product-specific subsidies and measures aimed at controlling production (quotas, compulsory fallowing, etc.). In this case, flexibility can be granted regarding the specific form of the policy at the national and regional levels (WB-BML, 1998, P. 23).

Transfer payments to agriculture motivated by distribution policy, however, are not the responsibility of the EU. Such payments must be oriented to the criteria which otherwise (outside the field of agriculture) prevail in the Community. These criteria are quite different amongst the Member States. For good reasons, the responsibility for organising the social security and tax systems lies essentially with the Member States. This should also be the case for regulations specific to agriculture in those areas.

A mixed position is occupied by those equalisation payments which are granted as price equalisation payments by EU reforms and Agenda 2000. As long as these can be viewed as an equalisation for the abrupt dismantling of EU price supports, and are for a clearly defined transitional period, they fit the category of production-oriented equalisation, which is to be laid down uniformly for all Member States in the interest of a common market and are logically administered at the EU level. Such payments are to be viewed quite differently, however, if they are to be granted for a long time. In that case, they require a different justification. If one wished to establish these as long-term transfer payments, motivated by distribution policy, then these would have to fit the national criteria which otherwise prevail in the particular local political environment. More consistently, the responsibilities for decision making and financing would have to be reallocated to the national level. The task of the EU would be to see to it that these payments are in accord with common regulations on competition. If one wanted to establish the payments as area-specific payments for services provided by agriculture which are not settled via the market for agricultural products, then one must determine whether these services are performed across borders or whether they are mainly regional. Only in the cross-border case should the responsibility for such payments be reallocated to the EU.

Likewise, in agricultural and rural structural policy and in agro-environmental policy, the decision-making responsibility of the EU would have to be limited to global problems extending beyond the borders of the Member States. Otherwise the decision-making responsibility would be assigned to the national level. The EU should define the framework, which would then be put in a concrete form and implemented by Member States. At the same time, the EU should, however, also more closely observe its control function. It should see to it that the national activities in these areas do not contravene the common regulations on competition.

The solution of problems in structural and environmental policy with mainly regional reference should also be anchored at the level of the Member States, particularly decision-making, implementation, and financial responsibilities. The payments to farmers for ecological efforts, already mentioned, are of this province. Depending on the vertical structure of the individual Member States, the regional/local levels should be included to a greater extent in the principle of subsidiarity. In the FRG, a reform of the joint tasks 'improvement of the agricultural structure and of coastal protection' and 'improvement of the regional economic structure' should bring about a clear demarcation of responsibilities between the federal government and the federal states. At the least, those measures that have no effects across borders should be assigned as the exclusive responsibility of the federal states (WB-BML, 1998, P. 26 ff.).

EU agricultural policy must be relieved from having to effect specific public finance transfers between the Member States. This function can, in so far as is necessary, be more efficiently performed by general cohesion funds, amongst others.

Such a redistribution of responsibilities can be advanced not only for agricultural policy. It must be part of a fundamental reform of the system for distributing tasks and financing, both at the EU level, and in the FRG. Such fundamental debate about reform will take place. In this debate, the problems of rural areas will require greater consideration.

Final Considerations

Against the background of changed underlying conditions for integration and increased integration of European agriculture in world trade, the extensive transfer of agricultural policy responsibilities to the EC is in need of revision. In particular, the opening of the EU towards Central and Eastern Europe also requires a reconsideration of the mechanisms of EU agricultural policy and their incorporation in a comprehensively planned structural policy. With the Agenda 2000 reforms, the EU should be made truly 'ready for accession'. The CAP should be so reformed that it can continue to exist in an enlarged Community. There is still a long way to go before this objective will be achieved; in the Agenda 2000 negotiations agreements have been made only in relation to immediately impending problems.

When the EU admits the Central and East European states, the spectrum of national expectations and interests in comparison with present Members will be expanded to an extreme degree. One cannot imagine that the 'acquis communautaire' would be applicable to an EU expanded by the states which have made application for admission (Lammers, 1999). Thus far there is a chance that, in the course of negotiations for producing an EU 'ready for expansion', some of the previously discussed reform ideas regarding will produce a breakthrough. The history of integration up until now has shown, however, that economic institutional approaches for distributing responsibility have little power in the face of the politico-economic circumstances prevailing when integration actually takes place. Nevertheless, such approaches can help to explain problem areas and point the way for pending political decisions.

References

Agrarbericht (1998). Agrar- und ernährungspolitischer Bericht der Bundesregierung [Report on Agricultural and Nutricional Policy of the Federal Government], Bonn.

BMF (1999a) [Bundesministerium der Finanzen−Federal Ministry of Finance]. Die Steuereinnahmen des Bundes und der Länder im Haushaltsjahr 1998 [Tax Revenue of the Republic and the Federal States in the Fiscal Year 1998], Bonn.

BMF (1999b) [Bundesministerium der Finanzen=Federal Ministry of Finance]. Schätzvorschlag des Bundesministeriums der Finanzen für den Arbeitskreis Steuerschätzung [Proposal of the Federal Ministry of Finance for the Working Group Tax Estimation], Bonn.

Deutsche Bundesbank (ed.) (1999). Monatsbericht Juli 1999 [Monthly Report July 1999], 51. Jg., No. 7.

Henrichsmeyer, Wilhelm / Witzke, Heinz Peter (1994). Bewertung und Willensbildung [Assessment and Intention Development], *Agrarpolitik*, vol. 2, Ulmer, Stuttgart, P. 542 ff.

Koester, U.(1999). The Role of Germany in the Common Agricultural Policy, in this volume

Lammers, Konrad (1999). Europäische Integration und räumliche Entwicklungsprozesse: Wo bleibt die nationale Ebene? [European Integration and Spatial Development Processes: Where Remains the National Level?] *HWWA-Diskussionspapier*, No. 75, HWWA-Institut für Wirtschaftsforschung, Hamburg.

Lehner, Franz (1979). Politikverflechtung: Institutionelle Eigendynamik und politische Kontrolle [Policy Interweavement: Institutional Momentum and Political Control]. In: Joachim Matthes (ed.), *Sozialer Wandel in Westeuropa. Verhandlungen des 19 Deutschen Soziologentages*, 611-625, Campus-Verlag, Frankfurt/Main.

Niehaus, H. (1957). Leitbilder der Wirtschafts- und Agrarpolitik in der modernen Gesellschaft [Models for Economic and Agricultural Policy in Modern Society], Stuttgart.

Plankl, R. (1996). Die Entwicklung des Finanzmitteleinsatzes für die Förderung umweltgerechter landwirtschaftlicher Produktionsverfahren in der BR Deutschland [Development of Expenses for the Promotion of Environmentally Friendly Agricultural Production Processes in the Federal Republic of Germany]. Agrarwirtschaft, Frankfurt/Main 45, 6, P. 233-239.

SRW (1980) [Sachverständigenrat zur Begutachtung der gesamtwirtschaftlichen Entwicklung=Advisory Council for the Appraisement of the Economic Development in Germany] (ed.): Unter Anpassungszwang, *Jahresgutachten 1980/81 [Yearly Report 1980/81]*, Verlag W. Kohlhammer, Stuttgart und Mainz, P. 176 ff.

SRW (1998) [Sachverständigenrat zur Begutachtung der gesamtwirtschaftlichen Entwicklung= Advisory Council for the Appraisement of the Economic Development in Germany] (ed.): Vor weitreichenden Entscheidungen, *Jahresgutachten 1998/99 [Yearly Report 1998/99]*, Metzler-Poeschel, Stuttgart.

Scharpf, Fritz W. (1994). Optionen des Förderalismus in Deutschland und Europa [Options for Federalism in Germany and Europe], *Theorie und Gesellschaft*, vol. 31, Campus Verlag, Frankfurt/Main, P. 16.

Scharpf, Fritz W. (1977). Politische Bedingungen einer aktiven Agrarstrukturpolitik [Political Conditions for an Active Structural Policy in Agriculture]; WZB-discussionspaper MDP 77-101, Berlin

Statistisches Bundesamt (1999). Statistisches Jahrbuch 1999 [Statistical Yearbook 1999], Metzler-Poeschel, Stuttgart.

Strauß, Franz Josef (1969). Die Finanzverfassung [The Financial Constitution], *Geschichte und Staat*, vol. 144/145, Günter Olzog Verlag, München.

Wiedman, Th., (1996). Idee und Gestalt der Regionen in Europa [Idea and Shape of the Regions in Europe], Baden-Baden.

WB-BML (1998) [Wissenschaftlicher Beirat beim Bundesministerium für Ernährung, Landwirtschaft und Forsten=Scientific Advisory Board at the Federal Ministry for Food, Agriculture, and Forestry]: Kompetenzverteilung für die Agrarpolitik in der EU [Distribution of Competences for Agricultural Policy in the EU], *Angewandte Wissenschaften*, No. 468, Bonn.

WB-BMWi (1999) [Wissenschaftlicher Beirat beim Bundesministerium für Wirtschaft und Technologie= Scientific Advisory Board at the Federal Ministry for Trade and Industry]: Neuordnung des Finanzierungssystems der Europäischen Gemeinschaft, *Stellungnahme* [Reorganization of the Financial System of the European Community, *Experts' Opinion*].

Zimmermann, Horst (1990). Gewichtsverlagerung im förderativen Staatsaufbau unter EG-Einfluß? [Shifts in the Federative Structure Under the Influence of the EC?] Wirtschaftsdienst, 70. Jg., P. 451-456.

Chapter 10

The Role of Germany in the Common Agricultural Policy

Ulrich Koester
Institute of Agricultural Economics, University of Kiel[1]

Introduction

At the time of its birth, the European Union's Common Agricultural Policy (CAP) was supposed to be a major engine of European market integration.[2] There was a widely held hope that positive integration in agriculture would force other sectors to follow the same route. However, expectations have not materialized. The annual price negotiations for agricultural products made evident the divergence in the national interests of EU Member States. Decisions were dominated by compromises between member countries rather than EU-wide interests.

The aim of this chapter is to highlight the German influence on the birth and evolution of the CAP. The chapter starts with a stylized decision-making model which provides for the structure of the subsequent analysis. To understand policy formation in the EU, as in any other country, one must analyze the political market. The influence of a member country can only be exceptional if its domestic political market differs from that in other countries, and if the institutional framework of decision making on the EU level allows for pursuing a particular national interest. Hence, a short section will explore the specifics of the German political market for agricultural protection. The chapter is based on the perception that poli-

---

[1] The author acknowledges helpful comments by S. Tangermann, C. Weiss, and B. Brümmer.
[2] The more recent term European Union (EU) will be used throughout this chapter even if reference is made to historical periods when the terms European Economic Community or European Community were in use.

cies are highly path dependent, and that from the start the CAP mirrored specific national interests. It will be shown how German policy makers managed to shape the CAP in their interests. The main sections of the chapter analyze how Germany could influence major changes in the CAP. It is argued in the last section that the attitude of the present German government and prevailing external pressures (WTO, eastward enlargement) will likely reduce Germany's influence on further evolution of the CAP.

A Model of the Decision-making Process

Prescriptive policy decision models assume that policy makers try to maximize a well-defined objective function taking economic conditions as a given (Streit, 1991,316). It would follow that policies change because objectives and/or economic conditions change. This prescriptive decision-making model does not reflect reality in a reasonable way, particularly as it does not take into account that present choices greatly depend on past decisions. In actuality there is a certain degree of path dependency in policies (Figure 1). Path dependency explains why economic policies change only gradually over time, 'ratcheting' their effects, unless significant shocks in the economic environment force massive adjustments (like the Great Depression in the 1930s).

Following this basic idea, the status quo of a given set of policies determines the policy decisions in the next period. This assumption is in contrast to the prescriptive policy approach where policy makers are seen as newly born in each period and not bound by past decisions or conditions on the political market. According to the underlying assumption in this chapter, the specific economic and political environment at the birth of the CAP affects the evolution of the policy in the following years and also determines the influence of member countries. Hence, if one wants to identify the national influence of a specific member country, it is worthwhile to first investigate what influence this member country had on the design of the CAP at the time of its inception. This approach seems of special interest if the country under consideration, Germany, was one of the founding members of the European Community. Second, it is worthwhile to consider those EU institutions and organizations which may have allowed individual countries to exert special influence on decisions. Third, it may be of interest to determine characteristics of a country which may lead to a special interest in EU policy making.

Figure 1:

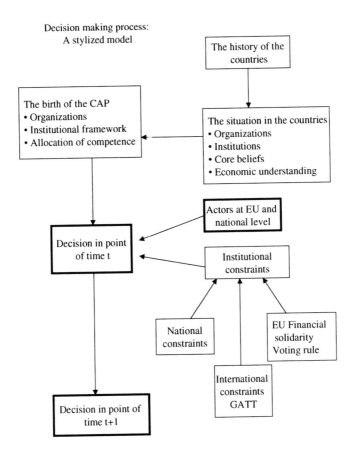

In searching for answers to these questions, special emphasis will be put on the effects of institutions and organizations. These two terms are used according to the definition of North (1996): institutions are the rules of the game which constrain the behavior of agents and make their actions predictable; organizations are groups of individuals bound by some common purpose to achieve objectives.

The Specifics of the German Political Market for Agricultural Protection

At the time of the CAP's inception, the political market for agricultural protection in Germany had some specifics founding the CAP which differed from those in other EU countries. First, there was only one highly organized farmers' union, while other countries had more than one, often with conflicting interests (Katranidis 1985; Mahlau 1985; Mahlau 1985; Matthews and Trede 1983; Matthews and Trede 1983; Meyer 1983; Meyer 1983; Trede 1983; Trede 1983; Trede and Filter 1983; Treiber 1983). Hence, farmers in Germany could exert strong political pressure. Second, that influence was probably more effective than in other countries due to two features of the German political system: the political party system and German federalism. Tangermann (1993) highlighted the specifics of the German party system and the importance of farmers in the outcome of any election and even in the formation of the government. Until 1998 the political influence of farmers was much stronger than one would have expected if their share in the labor force were the only yardstick. Federalism in Germany further strengthened the influence of the agriculturists as hardly a year passed without a state election resulting in the federal government somewhat adjusting its policies each time to improve the electoral chances of the respective party. Third, in contrast to other countries, especially the UK, there was never a strong political countervailing power balancing the influence of farmers in Germany. Consumers were never strongly organized and had other priorities than influencing food prices. Industrialists were quiet concerning agricultural protectionism as some of their national organizations were interested in agricultural protectionism and others did not regard it as an issue. The economic environment, with fairly stable consumer and food prices, did not cause a macroeconomic stabilization problem as in France, Italy, or the UK. Hence, the market structure in Germany for agricultural protectionism differed from other member countries.

Germany's Role in Shaping the CAP

The birth of the CAP caused severe pain for the founding members of the EU. Integrating the agricultural sectors was much more difficult than integrating the manufactured goods sector where the main task was the reduction of trade barriers. In contrast, integrating the agricultural sectors demanded positive integration. The EU members had to agree on harmonizing their national policies. The agreement to establish one common market with market organizations for the individual agricultural commodities required an agreement on a common price level, the form of border protection, finances, and on the rules for linking exchange rate variations to changes in national and common price levels. Germany had, due to its agricultural policy at the time and the German political market for agricultural protection, a specific national interest in solving these questions.

The German Interest in Setting the Common Price Level

High agricultural protection in Germany followed from crucial decisions made in the 1950s. When the German government generally accepted the Wirtschaftsordnung of the 'Soziale Marktwirtschaft' as the general principle for the economy in 1948, exceptions were allowed for agriculture. Basic decisions for the agricultural sector in Germany were made in the period 1949 to 1951. The Government had asked advisers for recommendations on how to treat the agricultural sector (Kluge 1989). The advisory committee consisted of two opposing groups. Group A, the most well-known agricultural economists of that time, favored the general acceptance of a competitive market economy for the agricultural sector, although they suggested some border measures for a transition period. They strongly objected to internal price regulations and, above all, they advocated the abolishment of the quota system for sugar production. In contrast, group B, headed by President Fritz Baade of the Institute of World Economics and the vice president of the same institute, together with some representatives of the agricultural administration and agricultural associations, argued for a highly regulated sector. The argument was not, at least explicitly, that food security was at stake. This argument might have been close at hand because of the famine in Germany during the second half of the 1940s. Amazingly, the argument of this group of advisers was that German farmers were not competitive at prevailing world market prices and hence, income support through price protection should be provided. It should be added that this view was widely shared by government officials and likely expressed the core beliefs of German society at that time. In 1951 Chancellor Adenauer stated:

> I have always been a friend of agriculture and I am deeply convinced that the preservation of a sound nation of farmers is one of the fundamental tasks of reasonable statesmanship. I am convinced that the level of agricultural prices has to be according to my conviction at par with prices in the overall economy. The Federal Government will implement suitable measures to secure an agricultural price level which covers actual production costs. . . We are completely certain that it is best to promote domestic production in order to reduce imports instead of promoting exports at any price. . . By promoting domestic production and, thus, reducing imports we will be less dependent and more stable. That is in the interest of consolidating our political system and can be considered as an extraordinary success (Puvogel 1957, 30, translated from the German by the author)

The quotations reveal that the German government considered farming not just one among many different economic activities, but highly important for society at large. Understanding of the highest German official seemed to be out of line with current economic thought. Even at that time agricultural economists widely agreed that price setting based on cost of production causes significant problems and would distort markets and moreover, that import substitution was not a first best policy.

It was no surprise that the German farmers' union proposed legislation that laid the legal foundation for permanent protection of the agricultural sector. The discussion around this proposal and the final agreement on the law by Parliament illustrates the political market in Germany and hints at the influence of Germany on the formation of the CAP. The draft of the law by the farmers' union was accepted by the Christian Democratic Party with few modifications. The other parties generally supported the law, but the Socialist Democratic Party (SPD) and the German Party (DP) wanted even more specific commitments to agricultural support by the Government. The federal association of German industry stated that agricultural income should grow parallel with that in other sectors. This attitude on the part of industry might be surprising. However, one must take into consideration that the industry association comprises members, those producing agricultural inputs, with a strong interest in a prosperous agricultural sector. Hence, it was difficult for the association to speak out against the strong interests of some of its members. The economic and political climate and possibly the lively memory of starvation in the minds of the German population eventually led to the acceptance of the Agricultural Act of 1955 which can be seen as a cornerstone of German agricultural policy. The Act commits the Government to policy implementation compensating agriculture for existing natural and economic disadvantages as compared to other sectors (Article 1 of the Agricultural Act). The Act also called for an annual report to be prepared by the Government describing the income situation of farmers and comparing it with nonfarm incomes. It should be noted that a large majority of the German Parliament agreed to the diagnosis and the prescription in the Act. It can be assumed that this view was also held by the majority of German society.

Since the same German government that supported the Act had to negotiate the principles of the CAP only a few years later, it was understandable that the same line of reasoning was used: first, market forces lead to farm incomes lower than those in other sectors of the economy. Second, the Government has a commitment to support farm income. Third, the basis for comparing farmers' income is not total income of the farmer, but instead income from farming; the basic idea was that farmers (and their families) worked full time on the farm and had no additional income. Fourth, support policy can actually adjust farm income to that of nonfarmers.

It is quite obvious that the perception behind the Agricultural Act and the German negotiation strategy in the EU was not economically sound. First, there are some convincing arguments that the farm sector does not suffer more than other sectors from adverse natural and economic conditions. Second, the method used to compare income from farming with nonfarm income may lead to very strange results. It may well be that the calculations reveal a disparity (income in agriculture smaller than in non-agriculture) but nevertheless total income generated from farming might be far above average (see Koester 1992, for the method of calculation). Third, many farms have income from several sources, either because they farm part time or because they have capital or land income and hence, they may be quite happy as farmers even if their income from farming is lower than that of those employed in other sectors. Fourth, there is evidence that per capita income from farming cannot be raised by income support in the long-run; integrated factor markets will adjust income to opportunity costs, and labor income in agriculture cannot be increased by income support. Hence, policies which try to conserve the structure can hardly reduce the disparity.

The specifics of the German situation led to agricultural prices in Germany in the pre-EU phase that were higher than in the other founding member countries. It became clear in the negotiation of the CAP that of the various agricultural prices, cereal price levels were the most important. Germany and Italy had the highest cereal prices, but since cereal production is a much less important source of income in Italy than in Germany, there was little support for setting prices at high levels. Hence, Germany had to accept a reduced price level. Had the GATT rules been applied, which state that the average rate of protection of a new established customs union should not be higher than the average protection rates of the member countries in the pre-union period, a clear solution would have been found. However, other GATT member countries seemed to acknowledge the difficulties of following that rule. Most important, the USA, the main player on agricultural world markets, had great political interest in a politically viable EU. As a consequence, the German influence prevailed resulting in a relatively high external protection rate for agricultural products. Moreover, the Germans succeeded in transferring German style market organizations to the European level. The sugar market organization with a quota regime was a case in point; it was transferred to the EU level in spite of the conflict between the concept of production quotas and the basic idea of a customs union.

Germany's Influence on the Types of External Protection

The design of the European market organization required decisions on the type of external protection. Germany's domestic market organization for most products was based on an annual production/consumption balance sheet and contained quantitative import quotas and domestic price setting (Puvogel 1957). Ad valorum tariffs were out of question for Germany. Even the decision of the European Council to introduce variable import levies without using any import restrictions was hardly acceptable from the German position. The Germans finally agreed, but only because they were allowed to apply a safeguard clause to restrict imports in cases of internal market disturbance (Tracy 1989). Again it was the German legacy which hindered a more liberal trade regime and gave rise to the variable levy system used in the Community and remained unmodified up to the Uruguay Round. Evolution of the CAP proved that the decision on the variables levy system was most crucial.

The Role of Germany in the CAP

The Type of External Protection as Determinant of National Influence

Under the variable levy system, the Council of farm ministers used to annually set threshold prices, intervention prices, and target prices for the most important agricultural products. Hence, each year individual countries could exert their interests in the Council. Germany used this opportunity and pursued its farmers' interests at the EU level. In most cases the German minister of agriculture was very close to the farmers' union and he generally tried to pursue the interest of the farming population. Of course, in principle the German minister for food, agriculture and forestry is charged with serving the interests of the general population, but that was rarely enforced. It was always quite clear that the German minister had to fight for the well-being of farmers. The decision-making process in the Council is actually favorable for pursuing sectoral interests. Meetings are not open to the public, in contrast to legislative bodies in all the member countries, and the behavior of the individual Council members will likely not be made public. Hence, the individual minister could always pretend publicly that he had tried to do the best for society, but was forced to compromise so that a common decision could be reached.

The power of the individual Council ministers greatly depends on the voting rule in effect and on the information used for reaching decisions. The Treaty of Rome (1957), the constitution of the Community, foresaw majority voting as the rule and unanimity as the exception. However, in 1965, even before the first market organization came into force, France succeeded in applying unanimity voting whenever a country claimed that a vital national interest was at stake. This voting rule strengthened the influence of individual Member States, and Germany made full use of this opportunity. Based on its legacy, Germany steadily pushed for a higher protection rate. As one outstanding example, in 1985 the Council tried to agree on a reduction of grain prices by no more than 1.6 percent, the German minister did not agree, blocking a decision on grain prices.

The voting rule was officially changed in 1993 with the Maastricht Treaty. However, the informal agreement of the 1965 Luxembourg compromise is still at work: simple majority voting is still very rare and qualified majority voting is also the exception, applied in about 10 percent of the cases (Wessels 1996). The use of majority voting is gradually increasing, but in many cases there are still long negotiations to reach unanimous decisions. It is quite clear that this voting procedure allows individual countries to exert a strong influence on common decisions.

It is noteworthy that the Commission and the Council accepted the view that price changes should allow for income parity in spite of the huge variation of farm income among different countries, even if this perception was not backed by the formulated objectives in the Treaty. The German view seemed to have dominated. The architecture of EU organizations and institutions have most likely helped strengthen the German position.

One could expect that Germany would have been constrained by national expenditure caused by the increase in the agricultural price level. It should be noted that the implemented principle of 'financial solidarity' gives rise to flows of real income between member countries as a consequence of price changes (Koester 1977) and hence somewhat increasing the diverse national interests of the countries. Germany was among those countries which had to accept the highest loss in real income resulting from price increases; price increases contributed to higher expenditure for export subsidies due to a higher price gap between domestic and world market prices and boosted domestic production and thus, exportable surpluses. Germany had to pay the highest share of the increase in expenditure, becoming the 'paymaster of Europe'. Moreover, previously Germany was an importer of most agricultural products so increases in domestic prices directly taxed domestic consumers in favor of producers in other EU countries. Nevertheless, the German farm minister pushed for price increases. This clearly supports the view that the minister was not interested in the welfare of the general population, but only in the (perceived) welfare of farmers. Lack of transparency in the decision-making process and ignorance of its effects by the majority of the population have facilitated the pursuit of vested interests on the Community level.

The German Influence on the Development of the CAP

The CAP has changed over time and there is little doubt that some countries were in favor of a more liberal CAP while others pushed for more intense regulation. The driving forces for change in the CAP were budgetary pressure (the dominant influence up to 1992) and foreign trade restrictions. In this section the German influence on the most important changes in the CAP is discussed (see also Tangermann 1992, 1993).

The Role of Germany in the CAP

The Introduction of Monetary Compensatory Amounts in 1969

The decision to institute market organizations as the basic principle of the CAP demanded common institutional prices. Hence, a rule had to be established which allowed the conversion of these prices into national currencies (See von Cramon-Taubadel 1994). The founders of the market organizations agreed on one specific rule: common prices were set in units of account and converted to national currencies at the prevailing exchange rate. However, when exchange rates were adjusted, the rule implied that institutional prices of countries which devalued (revalued) their currency were increased (decreased) in national currency. Thus, there was a strict connection between exchange rate policy and agricultural price policy. It might well be that the founders of the agricultural market regulations had not imagined the monetary instability which later arose. Up to that time, exchange rates had been fairly stable and there was hope that they would be further stabilized by market integration.

However, the experience did not meet expectations. Germany had to revalue its currency in 1969 – only one year after having introduced the first common market organizations – and France had to devalue its currency. The German Government feared the protest of farmers and demanded a waiver. The French Government supported the German view as it was interested in avoiding price increases for farm products which had spurred inflation. Therefore, the Council agreed to introduce separate agricultural exchange rates ('green money') and related monetary compensatory amounts, which allowed agricultural prices to be shielded from changes in general exchange rates. The consequence was that there was no longer a common price level. Agricultural products crossing borders of EU member countries were either taxed or subsidized to make up for the difference between the agricultural exchange rate and the general exchange rate. This was a clear contradiction to the principles of a common market. As Germany periodically revalued it currency, its agricultural prices were continuously above those in most other Member States. This system of national segmented markets existed up to January 1993. Thus, the German preference for a hard currency (low domestic rate of inflation) and strong preference for agricultural price support undermined the principles of a customs union. Actually, agricultural markets became even less integrated than they had been before the introduction of the CAP (Koester 1984). The introduction of the borderless Single Market in 1993 did not allow control of product flows across EU member borders and hence, the system of 'green money' had to be abandoned.

The Introduction of the Switch Over System in 1984

Under the 'green money' regime, German institutional prices always had to be lowered after a revaluation of the German mark, unless the decision was made to keep the agricultural exchange rate unchanged. Due to financial pressure, the Commission and nearly all Council members called for an automatic alignment of the German 'green' exchange rate to the general exchange rate within a short period of time (Kiechle 1985). German minister of agriculture Kiechle however, opposed the resulting decrease of agricultural prices in Germany. Instead of leaving agricultural prices set by the Community unchanged and adjusting national prices in domestic currencies, in 1984 he proposed the switch-over system. Prices fixed in units of account were adjusted to avoid a decline in prices in a currency which had been devalued. As the German and Dutch currencies were the only ones which were revalued from time to time, the new system was clearly in favor of farmers in these countries. Any new currency appreciation resulted in a price rise in units of account. The consequence was of course, an increase in the rate of external protection following appreciations of the German mark. Hence, this new rule contributed to the pressure to change the CAP in later years.

The Introduction of the Milk Quotas System

Another occasion on which Germany's influence on the CAP became very obvious was the 1984 introduction of milk quotas. As observers noted, 'the position taken by the German Government in the political debate and policy-making process during the introduction of milk quotas are hardly understandable, if one judges the actions from the viewpoint of general "national interests" and the general philosophy of economic policy in Germany' (Petit et al. 1987, 53). The puzzle can be explained by the dominance of Minister Kiechle. His basic idea was, 'Quantities down, prices up.' Hence, he was the main proponent of solving the milk market problem ('butter mountains' and high budget outlays for export restitutions, domestic consumption subsidies, and storage) by introducing a quota system for milk producers which was accepted by the Council in

1984. This regulation was again in strong conflict with the principles of a common market. Reallocation of production was not allowed to those regions and farms that had a comparative advantage.

The experience with the milk quota system supports expectations agricultural economists had at the outset of the scheme. Cuts in production did not automatically lead to higher prices as there was still an exportable surplus and hence, domestic prices depended on price decisions made by the Council as in the previous period. It may be argued that the Council was able to set higher prices as production was curtailed by the quota system. However, price increases still boosted surpluses as demand declined. The demand effect became even stronger when the European Court of Justice decided that milk substitutes had to be accepted on all EU markets.

The introduction of the milk quota system was an expression of the German tendency to solve urgent problems piece by piece. It was overlooked that the milk quota would eventually have spillover effects on other markets and would require further supply management. Hence, it was no surprise that Germany advocated a voluntary set-aside program to solve the surplus problems on the grain markets.

The Introduction of the Set-aside Program

The voluntary paid set-aside program was introduced in 1988. Minister Kiechle suggested that cuts in grain production would allow institutional prices on EU markets to be set higher. Yet it was hard to convince the other members in the Council. Finally, they agreed on the set-aside scheme but allowed member countries quite some leeway in implementation. Indeed given the system of common finances, individual countries were well advised to make it unattractive for their farmers as it would cause a loss in welfare and incur additional national outlays (Koester 1989). Germany was the only country that provided generous incentives to its farmers who indeed took advantage of the program. However, it was quite costly for the country, both in economic and financial terms (Koester 1989; Poggensee 1993).

The voluntary set-aside program, introduced under German influence, was again a serious violation of the principles of a common market. An important factor of production was not used, and allocation of resources deteriorated in the Union.

The 1992 CAP reform, German unification, and the changing attitude of German industry

In 1992 the CAP was considerably reformed. The most important element of the reform package was a reduction in grain intervention prices by 30 percent. Such a significant change was unthinkable prior to that time. Up to the mid-1980s there was actually a general understanding among Council members that any price decision in the annual round should not lead to a reduction in nominal prices in any country. One should recall that Germany did not accept a reduction in German grain prices by 1.6 percent in 1985, the first time the gentlemen's agreement not to lower nominal prices in any member country was broken. Hence, it needs to be explained why the German agricultural minister in 1992, still Mr. Kiechle, who had fought for strong regulations in the past, finally agreed to the reform package. This about face is explained by two major changes in the political and economic environment in Germany: German unification and the attitude of industry (Cramon-Taubadel 1993).

It should be remembered that the family farm played a major role in German policy making, an important element of the core belief that the family farm is the most desirable and efficient farm organization. Indeed, having family farms had intrinsic value, contributing to specific social objectives in society. Hence, policy aimed at conserving as many family farms as possible. This perception of farming changed suddenly after German unification.

At the beginning of the transformation, the German government thought that the superiority of family farms would be reflected in rapid changes in the farm structure. As a result the German government directed more subsidies towards family farms than other agricultural organizational forms. However, the evolution did not meet expectations. Seven years after unification single-owner farms[3] (including part-time farms of over 1 ha) cultivated less than 20 percent of the arable land, and corporate entities about 60 percent. Farms in the former GDR are larger than those in the former GDR (Forstner and Isermeyer 1998). In 1994/95 single-owner farms averaged less than 40 ha in the west, but 160 ha in the east; the average size of corporate farms in the east was 1,721 ha that same

---

[3] Single-owner farms in the former GDR are generally not comparable to family farms in the other part of Germany. Family farms in the west own most of the land they cultivate, single-owner farms in the east have a high proportion of rented land.

year. On average eastern farms rent 90 percent of the land they cultivate in contrast to about 40 percent in west Germany.

This reality strongly conflicted with the perceptions of German policy makers. Hence, the core belief 'family farms are the best and natural organizational form of agricultural production' quickly changed. Many large farms in the east were highly profitable after a few years of adjustment. Moreover, the adjustment pressure exerted on eastern farms after unification was much higher than what was considered socially acceptable for farms in the west. Therefore, the German government weakened its request to conserve small-scale agriculture based on family farms. This may well have contributed to its willingness to accept CAP reform in 1992.

German industry changed its position shortly after the start of the Uruguay Round. Up to then the heterogeneous interests of the industry association had not taken a clear position against agricultural protection. However, the Federal Association of German Industry published a booklet in 1987 demanding a market reform (Bundesverband der Deutschen Industrie 1987). One may wonder what caused the change in the attitude of the Association. The foreword of the booklet said first, that an efficient agriculture is in the interest of industry as agriculture is an important supplier of inputs to the industry and an important buyer of industrial goods. It also stated that the international reactions against the protectionist CAP cause damage for the whole economy, and that the CAP was a burden to the European dialogue and might hinder the creation of a Single Market. Finally the Association reported that the CAP was not compatible with the philosophy of the 'Soziale Marktwirtschaft'. The same arguments had of course, been well propagated by economists for a long time and one may wonder why the Association of Industry expressed this view at that point of time.

There is a strong presumption that international pressure was the driving force. The Association was aware that ongoing GATT negotiations could come to an acceptable agreement if and only if the CAP was significantly changed (Cramon-Taubadel 1993). Thus, the German farmers' union had lost one of the most important quiet supporters of past policies. This helps to explain the position of the German government; it finally agreed to intervention price reduction in 1992, but succeeded in getting an agreement of full compensation for any income losses incurred as a result.

It should be noted that the 1992 CAP reform boosted budget outlays on the Community level. Expenditure was as the driving force for change in agricultural policies for a long time (Petit 1989). The 1992 CAP reform gave rise to a new era in which the major incentive for policy change was international pressure, specifically international trade commitments (Tangermann 1998).

The Agenda 2000 Package

The Agenda 2000 CAP reform was another important step. The early discussions on reform revealed the strong influence of the Commission. The Commission was firm that further reform of the CAP was needed to comply with the commitments of the Uruguay Round agreement, to prepare the Union for eastward enlargement and the next round of WTO negotiations, to make the agricultural sector internationally competitive, and to complete the 1992 reform. The official German opinion, expressed by the agriculture minister at the time, Borchert, was that further reform was not needed and further price cuts with less than full compensation were unacceptable.

Before the final decision in March 1999 there was a new government in place in Germany, a coalition of the SPD and the Greens. The SPD, the larger coalition partner, had long advocated CAP reform and proposed direct payments as a substitute for price support.

The position of SPD has changed markedly over time. The party was traditionally more prone to planning and less market oriented. Hence, the party supported the Agricultural Act of 1955. If one considers the background of the members of the agricultural committee of the Federal Parliament as an indicator of the vested interest of a party in agricultural decisions, a significant change can be observed over time. Breitling (1955) investigated the composition of the first Federal Parliament and of the agricultural committee. He found that the SPD had a low proportion of members of the parliament who had an agricultural affiliation (only about 2 percent), but SPD members made up about 40 percent of the agricultural committee. The legislation prepared by the committee expressed a generally strong interest in agriculture. The composition of the third Parliament was similar. However, the picture has completely changed in to the thirteenth parliament (1998-2002); of the twelve SPD members on the committee, only one considers himself an agriculturist. Understandably, these committee members fight less for agricultural interests than did those of 1949. The change in the party's ideology is also reflected in more recent party programs. The 1987–1990 program of the SPD called for a fundamental reform of agricultural policy and favored direct income payments (SPD 1987). However, the party still rejected world market prices for agricultural products. This view had changed before the 1998 German elections. The party demanded a more drastic reform than was proposed by the Commission, a reduction of export subsidies, and less restrictive market organizations (SPD 1999). Direct payments were proposed to smooth the effects of income reduction.

Having experienced this shift on agricultural policy, the German government should have been quite happy with the main elements of the proposal. Indeed, official statements of the current government differ widely from that of the former government; they are more market oriented. Funke, the new minister of agriculture after the 1998 elections, was president of the Council and concluded the negotiations. It is likely that both the German presidency of the first half of 1999 and the new German government contributed to the final agreement, which was less market oriented than the Commission's proposal, but nevertheless was a major reform.

Summary

- On the German political market, agricultural protection was more favored than in the other founding member countries of the EU or those EU countries under the CAP. German farmers were better organized than those in the other countries: farmers' votes played a crucial role in federal and state elections, nonfarm organizations abstained from influencing policies, and core societal beliefs supported farm protection. The perception that the family farm is the 'natural' organizational form of farms and should be supported by state intervention was widely accepted, more so than in most other founding EU member countries. France, one of the major early players, was much less focused on conserving the family farm structure than was Germany. In contrast The Netherlands favored structural change in agriculture in order to be internationally competitive.
- Germany had a specific interest in the formation of CAP. German agricultural price levels were higher than in the other prominent founding

member countries (except for Italy), so political pressure persisted for a continuation of the high price level policy.
- Germany had the strong conviction that income parity should be achieved by government intervention (as laid down in the Agricultural Act of 1955) and that price support was the adequate means to achieve that objective. France, the main negotiating country during the birth of the CAP, and a major exporter of agricultural products, could expect to benefit from a high agricultural price level. Consumers in importing countries such as Germany, were taxed in favor of French farmers. Hence, there were no strong opponents to a high price policy from the CAP's inception.
- The institutional arrangements set up in the CAP on the one hand widened the divergence in interests, and on the other hand helped member countries fight successfully for their national interests. These institutions were primarily concerned with voting procedure and financing arrangements. Even though the financial arrangements of the Union made agricultural protection for Germany more costly compared to a situation with an isolated Germany, Germany continued to push for higher protection, indicating its strong preference for farm support.
- The significant divergence in national interest among the member countries showed up in specific policy changes made over time. The introduction of the 'green money' system allowed Germany to have a higher price level than its partners. Other countries, especially France and the UK, favored a lower agreed price level in order to support consumers and curb inflationary pressure.
- The increase in the intensity of regulation was predominantly favored and supported by the German government. The introduction of the milk quota scheme is a case in point. It reflected the German tendency to cut production and increase prices. The German influence on the introduction of the set-aside program is another important case where the German government exerted pressure to intensify regulations.
- The German position on the CAP changed in the late 1980s with the upcoming Uruguay Round, and in the 1990s with German unification. The former event brought a new player onto the field; international trading partners became stronger and rigorously pushed for change. Unification came as a jolt to the core beliefs of German agricultural policy makers. The concept of family farms being superior lost influence as most of the new farms in the reunified Germany did not fit this

theory. It was mainly due to these determinants that Germany accepted the 1992 CAP reform. However, it insisted on an increase in the density of regulations (set-aside, allocation of quotas to member countries for arable land which can be planted without sanctions). The implementation of the new oilseed regime in Germany was a specific violation of the principles of the Single Market.

- The German impact on the Agenda 2000 reform proposals was significantly affected by the change in government. The old government under Chancellor Helmut Kohl tried to block any change in the CAP, pretending that no change was needed. The new government, which took office in October 1998, had a difficult start. It gained the presidency in January 1999 and had to bring the negotiations to a close. The experience with the new government seems to mark a significant change in agricultural policy making in Germany. The influence of the farmers' union and of the ministry of agriculture seems to have declined and some changes have been introduced which require stronger adjustments for the agricultural sector than most other sectors. Hence, in the future it can be expected that Germany will push less for higher protection of agriculture than it has in the past.

References

Breitling, R. (1955). Die Verbände in der Bundesrepublik. Meisenheim am Glan.

Bundesverband der Deutschen Industrie e.V. (1987). Agrarpolitik. Denkanstöße für eine marktwirtschaftliche Reform. Köln.

Cramon-Taubadel von, S. (1994). The European Community Common Agricultural Policy and the Search for a Unit of Account. Oxford Agrarian Studies. 2:2 107–122.

Cramon-Taubadel von, S. (1993). The Reform of the CAP from a German Perspective. Journal of Agricultural Economics. 44:3, 394–409.

Forstner, B. and Isermeyer, F. (1998). Zwischenergebnisse zur Umstrukturierung der Landwirtschaft in den neuen Bundesländern. Berichte über Landwirtschaft. 76, 161–190.

Hansmeyer, K.-H. (1963). Finanzielle Staatshilfen für die Landwirtschaft. Tübingen.

Katranidis, S. (1985). Agrarpolitik und Agrarsektor in Griechenland. Agrarpolitische Länderberichte EG-Staaten, Bd.9, Kiel.

Kiechle, I. (1985). ...und grün bleibt unsere Zukunft. Stuttgart.

Kluge, U. (1989). Vierzig Jahre Agrarpolitik in der Bundesrepublik Deutschland. Band 1. Hamburg und Berlin.

Koester, U. (1977). The Redistributional Effects of the EC Common Agricultural Policy. European Review of Agricultural Economics. 4:4, 321-345.

Koester, U. (1984). The role of CAP in the process of European integration. European Review of Agricultural Economics. 11, 129-140.

Koester, U. (1992). Grundzüge der landwirtschaftlichen Marktlehre, WiSo-Kurzlehrbücher, München.

Koester, U. (1989). Financial Implications of the EC Set-aside Programme. Journal of Agricultural Economics. 40:2, 240-248.

Mahlau, M. (1985). Agrarpolitik und Agrarsektor in Spanien. Agrarpolitische Länderberichte EG-Staaten. 10. Kiel.

Mahlau, M. (1985). Agrarpolitik und Agrarsektor in Portugal. Agrarpolitische Länderberichte EG-Staaten. 11. Kiel.

Matthews, A. und Trede, K.-J. (1983). Agrarpolitik und Agrarsektor im Vereinigten Königreich. Agrarpolitische Länderberichte EG-Staaten. 2. Kiel.

Matthews, A. und Trede, K.-J. (1983). Agrarpolitik und Agrarsektor in Irland. Agrarpolitische Länderberichte EG-Staaten. 3. Kiel.

Meyer von, H. (1983). Agrarpolitik und Agrarsektor in Italien. Agrarpolitische Länderberichte EG-Staaten. 7. Kiel.

Meyer von, H. (1983). Agrarpolitik und Agrarsektor in Belgien und Luxemburg. Agrarpolitische Länderberichte EG-Staaten. 8. Kiel.

North, D.C. (1996). Institutions, Institutional Change and Economic Performance. Cambridge University Press.

SPD. (1987). Zukunft für alle - arbeiten für soziale Gerechtigkeit und Freiheit. Regierungsprogramm 1987 - 1990 der Sozialdemokratisachen Partei. Bonn.

SPD (1999). Die Reform der Agrarpolitik. Internet: http://www.spd.de/aktuell/ euro6_26.htm

Petit, M., Benedictis de, M., Britton, D., de Groot, M., Henrichsmeyer, W. and F. Lechi (1987). Agricultural Policy Formation in the European Community: The Birth of the Milk Quotas and the CAP Reform. Amsterdam.

Petit, M. (1989). Pressures on Europe's Common Agricultural Policy. International Food Policy Research Institute, Ecole Nationale Superieure des Sciences Agronomiques Appliquees. Washington, D.C.

Poggensee, K. (1993). Die Flächenstillegungsprogramme der Europäischen Gemeinschaft - Analyse und Beurteilung. Agrarwirtschaft. Sonderheft.139. Kiel.

Puvogel, C. (1957). Der Weg zum Landwirtschaftsgesetz, München.

Streit, M.(1991). Theorie der Wirtschaftspolitik. WiSo-Texte, 4. Auflage. Werner-Verlag, Düsseldorf.

Tangermann, S. (1998). An Ex-Post Review of the 1992 MacSharry Reform. In: Ingersent, K.A., A.J. Rayner, R.C. Hines (eds.), The Reform of the Common Agricultural Policy. Houndmills, London, New York (Macmillan).

Tangermann, S. (1993). Agricultural Protectionism and the Agricultural Lobby in Germany: International Implications. American Institute for Contemporary German Studies. The Johns Hopkins University Washington. Seminar papers, Number 6.

Tangermann, S. (1992). European Integration and the Common Agricultural Policy. In: Barfield, C. E. and Perlman, M. (ed.), Industry, Services, and Agriculture. The United States faces a United Europe. The AEI Press, Washington. 407-451.

Tangermann, S. (1979). Germany's role within the CAP: Domestic Problems in International Perspective. Journal of Agricultural Economics. 30, 241-257.

Tracy, M. (1989). Government and Agriculture in Western Europe 1880-1988.

Trede, K.-J. (1983). Agrarpolitik und Agrarsektor in den Niederlanden. Agrarpolitische Länderberichte EG-Staaten. 1. Kiel.

Trede, K.-J. (1983). Agrarpolitik und Agrarsektor in Dänemark. Agrarpolitische Länderberichte EG-Staaten. 4. Kiel.

Trede, K.-J. und Filter, W. (1983). Agrarpolitik und Agrarsektor in der Bundesrepublik Deutschland. Agrarpolitische Länderberichte EG-Staaten. 5. Kiel.

Treiber, W. (1983). Agrarpolitik und Agrarsektor in Frankreich. Agrarpolitische Länderberichte EG-Staaten. 6. Kiel.

Wessels, W. (1996). Europäische Union – Entwicklung eines politischen Systems. In: Ohr, R. (ed.), Europäische Integration. Stuttgart. 19-45.

Chapter 11
Agriculture and the Environment

Rainer Marggraf
Institute of Agricultural Economics, University of Göttingen

Introduction

As in many countries, the condition of the environment in rural areas in Germany is unsatisfactory. There is much talk about agricultural environmental problems and a general consensus that agriculture puts too much pressure on the environment and that not enough consideration is given to possible positive environmental effects. In economic terms this means that too many negative and too few positive external effects are associated with the production of agricultural market goods.

In this chapter, the extent to which German policy has reacted to this problem is presented. In addition, possibilities for further policy development with respect to market-oriented aspects are discussed. In the next section a short overview of the German agricultural environmental effects both past and present is given. The goal of agricultural policy is to control these effects by influencing agricultural land use. In the second section, the fundamental structure of agricultural environmental policy is explained. The policy makes use of legal and economic instruments which are discussed in sections three and four, respectively. The final section is dedicated to the important question of the extent to which agro-environmental policy can be redesigned to be more market oriented.

Environmental Effects of Agriculture

Middle Europe was dominated by forest ecosystems for hundreds of years. Agriculture has drastically and extensively changed the original natural landscape resulting in totally new ecosystems (Haber, 1985). The so-called agrarian revolution that took place in the Neolithic era, i.e. the transition from hunter-gatherers to a settled way of life, marks a critical turning point with regard to the use of natural ecosystems and the intensity of human intervention in the natural cycle/balance. During the time of the hunter-gatherers, environmental interventions occurred, however, they were generally local and often reversible (Remmert, 1988). The hunter-gatherer existence was not totally replaced during the period of innovation (the transition to crop cultivation, animal husbandry, and establishment of settlements), but it was permanently supplanted. The forest matrix largely remained up to the beginning of the Middle Ages. Not until several great clearing phases, starting in about 900 B.C., was the Middle European landscape fundamentally changed from forest to a primarily open, cultivated landscape.

This cultivated landscape, often described as 'extensive agriculture', was partly characterised by very intensive use of a particular area. However, as a rule, a well-structured mosaic landscape resulted due to limited technical options and the need for cultivation of a variety of crops (for self-sufficiency). This type of landscape was generally sufficient to link biotopes (Blab, 1992). Although imbalances and damage resulted locally, the spatial and temporal extent of human influence remained relatively limited. Human intervention in nature resulted in an increase in the variety of species and biotopes due to the wide spectrum of anthropogenic sites and the many gradients added to the existing natural and near natural ecosystems (Haber, 1972; Schumacher, 1997).

After 1850, human interventions had an increasingly negative effect on the ecosystems (Beck, 1996). Historically, marked agricultural intensification came with the advent of mineral fertilisers and implementation of fossil fuels. More and more powerful farm machinery required a well-developed network of paths and roads, fields adapted for the machines, and finally, the joining of scattered property (Haber, 1988). The reallocation of land in the 1950s was designed to 'clean up' the landscape. The ecological benefits of varied crop rotations became limited causing the need for more pesticides. Finally, the widespread use of fertilisers not only caused large-scale eutrophication of the landscape with diminished soil structure and quality, but also considerable groundwater pollution (Haber, 1988). Scientists unanimously believe that intensive agricultural practices are the primary cause of a decline in species (Bauer and Thielke, 1982; Blab and Kudrna, 1982; Korneck and Sukopp, 1988) and pollution of land and water.

Environmental effects of agriculture were occasionally discussed by natural and social scientists. In 1971 an independent advisory board was established in the Ministry of the Interior (since 1990 in the Ministry of the Environment). It is assigned the task of monitoring environmental quality and degradation, and assessing ways to avoid further damage. In 1978 the board concluded that German agriculture pollutes the environment. In 1985 the board presented a detailed report of the environmental effects of modern agriculture, thus opening a public and political discussion. By now, it is generally known that the negative environmental effects of the present intensive agriculture in Germany include the following problem areas:

- the decline in species and biotope variety (especially due to the damage, reduction, fragmentation and elimination of nature-oriented biotopes and landscape elements, the levelling, and grading of land detrimental selection effects of pesticides)

- the uniformity of the landscape (due, for example, to the elimination of landscape elements such as hedges and ponds in favour of larger fields)

- the pollution of groundwater by nitrates and pesticides
- the debilitation of soils due to structural damage, erosion, and the resulting pollution of substances
- surface water quality, particularly as a natural biotope and a carrier of plant nutrients and pesticide residues
- local air pollution and its contribution to global warming (for example, by the increase in methane and ammonia emission as a result of intensive livestock production)

Only two positive effects offset the negative environmental effects. The open landscape of Middle European agricultural cultivation and extensive land use contribute to half-natural biotopes which advance species' sustainment.

By now, broad consensus exists in society regarding the necessity for a reduction in environmental pollution. In addition, many recent willingness-to-pay studies conducted in Germany have documented that the level of these environmental effects is negatively evaluated based on efficiency as well. Therefore a reduction of soil, air, and water pollution and increased efforts in the maintenance of species, biotopes, and the landscape contribute to positive welfare effects.

General View on Agro-environmental Policy

Policies can govern the behaviour of individuals by direct regulation, economic incentives, and moral suasion (OECD, 1994). Agro-environmental policy uses all these approaches to influence farmers' behaviour.

Figure 1:
Governing of agricultural activities by agro-environmental policy

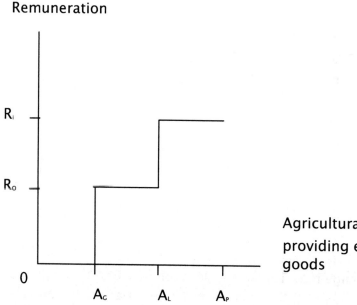

A_P = activities induced by agro-environmental programmes
A_L = activities laid down in law
A_G = good agricultural practice
R_I = remuneration covering opportunity costs and including an incentive component
R_O = remuneration covering opportunity costs

Agriculture and the Environment

Figure 1 shows agricultural activities providing positive contributions, i.e. activities that lead to a reduction of negative external effects (e.g. reduction of ground water pollution, soil erosion) or securing positive external effects (e.g. maintenance of the landscape).

In economic terms, things which are valued positively by individuals are called goods. According to this linguistic usage, the expression 'providing agro-environmental goods' is used in the following as a generic term for a reduction of negative external effects and a maintenance or intensification of positive external effects of agricultural production. Providing an agro-environmental good may require that certain activities are carried out, e.g. the maintenance of half-natural biotopes demands extensive agricultural use or its simulation. However, other agro-environmental goods are realised only when farmers refrain from certain activities. For example, quality improvement of the natural resources soil and water requires that farmers use fewer fertilisers and pesticides. Thus 'activity' can mean not only acting but also refraining from certain activities.

All activities that shall be governed by agro-environmental policy are marked A_p. These activities can be divided into mandatory (A_L) and voluntary activities ($A_p - A_L$). Many laws and regulations of agro-environmental policy as well as guidelines and recommendations refer to 'good agricultural practice' or 'orderly agriculture'. These activities (A_g) are a subset of the activities prescribed by law.

Good agricultural practice constitutes a duty for the farmer for which he can be 'rewarded' only by the market. Fulfilment of this duty and the resulting environmental benefits are considered by agro-environmental policy as activity not worthy of remuneration. The situation is different for the subset of activities laid down in law that exceed good agricultural practice ($A_L - A_g$). These activities must be provided by farmers who are under stricter requirements due to location, e.g. water protection areas, nature preserves. For that, they are to receive remuneration equal to their opportunity costs (R_O). Agro-environmental policy tries to influence activities that go beyond A_L with the help of agro-environmental programmes targeted at all farmers in Germany. Participation is optional, but made attractive by remuneration that exceeds opportunity costs (R_I). This economic incentive is supplemented by training programmes (which are part of the persuasive instruments) to familiarise farmers with environmentally acceptable production methods.

Due to the federal structure of Germany the sixteen Länder are primarily responsible for the concrete establishment of agro-environmental programmes. However, the Länder have different ideas about the concrete contents of a good agro-environmental policy, and conditions that vary considerably. For this reason the Länder neither promote the same activities nor do they calculate opportunity costs in equal ways or offer the same incentives.

The three above mentioned components of agro-environmental policy have varying application to agricultural land in Germany. As already mentioned, the rules of good agricultural practice must be upheld for all farmland. About 10% of farmland consists of water protection areas and nature preserves with stricter legal demands. Yet such demands do not lead to economic disadvantages for the farmers concerned. Agro-environmental programmes affect 30% of farmland, but proportions in the individual Länder vary considerably (Osterburg et al., 1997).

Agro-environmental Law

Agro-environmental law is constituted by specific regulations of agricultural law that are relevant only for farmers and by general regulations of environmental law. Here the term 'good agricultural practice' or 'orderly agriculture' is of major importance. It is supposed that good agricultural practice and protection of natural resources are identical goals. For example, the third amendment of the Federal Law on Nature Conservation of 1998 states that agricultural land use and forest economy generally do not contradict the aims and principles of protection of nature and conservation of the landscape.

The expression 'good agricultural practice' is an undetermined term that must be put in concrete forms for application (Winkler, 1994). Actually there is agreement only that agricultural use is good agricultural practice when pursued in accordance with the regulations on water, plant protection products, fertilisers, and disease control. Though only for a few activities have precise criteria been established for good agricultural practice, e.g. for fertiliser application quantities and dates.

The definition of good agricultural practice is partly very differentiated and bound to facts that farmers can often only understand with nearly unreasonable efforts and extraordinary know-how. Such regulations are not clear; where legislation exactly defines good agricultural practice, it tends toward overregulation. On the other hand, there are large areas of agricultural activities for which only vague ideas about good agricultural practice have been outlined, particularly for production techniques and cultivation methods that are environmentally friendly. As good agricultural practice is not yet clearly defined, it is not possible to precisely govern by law individual farmers' use of the environment. Symbol A_P in Figure 1 does not mark a well-defined set of agricultural activities.

Agro-environmental law generally only demands more than good agricultural practice from farmers who operate in water protection areas and/or other protected areas (Hötzel, 1996). Restrictions on land use, fertilisers, and/or plant protection are imposed on these farmers.

The German Federal Law on Water Resource Management, and since 1998 the Federal Law on Nature Conservation state that the economic losses caused by these restrictions must be compensated on the grounds that farms in protected areas can not be placed at a disadvantage against comparable farms outside protected areas. Details on the calculation and financing of compensation payments have to be stipulated in laws of the Länder. The Länder have fulfilled this obligation as far as compensation in water protection areas is concerned, but they have used quite different approaches.

For calculating compensation payments, most Länder have chosen a system taking into account different regions. There are three ways to finance compensation payments: a) direct beneficiaries pay compensation for the restrictions in the protected areas they use (decentralised regulation); b) the Länder finance compensation payments by imposing special rates (centralised regulation); or c) compensation is paid from Länder budgets, i.e. by taxes (Völsch, 1993). So far only the first two alternatives have been implemented. Experience in the different Länder has shown that decentralised water protection is advantageous (Kemper, 1992). In particular, such local water authorities having a greater responsibility and more direct contact with agriculture, have improved programme monitoring and the taking into consideration of regional characteristics. A decrease in administrative expenses can be realised when standardised contracts are developed which simplify and unify compensation procedures and avoid legal uncertainties. Since the costs for water authorities of compensation payments are passed on to the consumer, water prices depend on the size of the water protection area, the proportion of farmland in the protected area, and the volume of water involved. This has lead to divergent price developments for drinking water.

The 1998 Federal Law on Nature Conservation gave the Länder three years to implement compensation payments. The options for financing compensation payments, apart from taxes, are limited, although financial alternatives akin to those stated earlier could be taken into consideration. Implementation using funds collected from direct beneficiaries requires that beneficiaries be defined. Thus, individuals who use nature areas for recreation should pay an admission fee as they are receiving a recreational benefit. Collection of such fees, however, requires efficient exclusion mechanisms and appropriate size and quality of the nature area. When there are only few people for whom the nature reserve has a non-use value, this decentralised financial method would theoretically more closely correspond to the generally accepted principle of equivalence than would centralised financing by taxes. When realising this approach, it must also be noted that decentralised financing is entirely new for the German population and consequently will be unpopular. Therefore, at least in the initial years, sufficient revenues cannot be expected.

With regard to imposing special taxes for nature conservation, appropriate instruments can be established, e. g. a nature conservation tax levied on visitors of neighbouring regions by local authorities to finance nature reserves (such as a tax on tourist beaches). From the legal point of view, authorities may only demand a fee from visitors who actually enjoy a use value from the nature area. Consequently a nature conservation tax paid by every visitor would only be legal if the German legal system accepted the concept of a non-use value, as under American law (Brown, 1993).

Agro-environmental Programmes

German farmers were first offered participation in agro-environmental programmes in 1980 (Höll and von Meyer, 1996). These programmes were supposed to induce farmers to use natural resources more carefully than was legally prescribed. At first programmes were directed at particular aims and measures. Examples are the Bavarian programme for the protection of meadow birds and the programmes in North-Rhine-

Westphalia and Lower-Saxony to reduce distribution of semi-liquid manure (all established in 1984). Until 1992 the Länder programmes and their supplemental regional and local programmes were organised and financed largely independent from EC agricultural programmes. But in 1985 the EC offered Member States support to regional nature conservation programmes and extensification measures 'in ecologically particularly sensitive areas' by granting subsidies on a national level (Baldoch and Lowe, 1996). Since 1987 the EU co-finances programmes under certain conditions. As a result, the number of agro-environmental programmes increased to more than fifty by the end of the 1980s. During the CAP reform of 1992, Member States' agro-environmental programmes gained new status as part of the supporting measures. First, Member States were encouraged to establish agro-environmental programmes by VO 2078/92. Second, the EU programme financing, typically 50% (and in some less favoured areas up to 75%), is no longer administered by the Guidance Fund but by the Guarantee Fund of the EAGGF. In this context a financial limit was fixed for each Member State for the first five years. Financing by the Guarantee Fund is supposed to underline the permanence and validity of European agro-environmental policy.

Article 1 of the EU Regulation 2078/92 has three principal targets: 'A Community aid scheme ... is hereby instituted in order to accompany the changes to be introduced under the market organisation rules, contribute to the achievement of the Community's policy objectives regarding agriculture and the environment, contribute to providing an appropriate income for farmers'. The motivation for the Regulation is the establishment of '... an aid scheme to encourage substantial reduction in the use of fertilisers and plant protection products or the use of organic farming methods can help not only to reduce agricultural pollution but also to adapt a number of sectors to market requirements by encouraging less intensive production methods'.

Thus it becomes clear that the EU agro-environmental policy not only intends to encourage agriculture to provide environmental goods, but also to balance markets and guarantee adequate income for farmers. Aid schemes of Regulation 2078/92 which were developed on a national, regional, or partly local levels also consider these objectives.

When establishing agro-environmental programmes it must surely be taken into account that they not only affect the environment but also the economic system. But these 'side effects' are not to be seen as independent aims. The complementarity of 'provision of agro-environmental goods', the 'balance of markets', and of 'income maintenance for farmers' is to be questioned. Neither a permanent securing of incomes in agriculture, nor the balance of markets are legitimate political aims in a market economy. The first aim is not legitimate as there can be no guarantee of income exceeding the subsistence level. And a permanent national income maintenance policy would increase structural adjustment difficulties in agriculture. The second aim is not legitimate as balancing agricultural markets (not to be confused with stabilisation) has only become an issue since the market clearance function became invalid.

An assessment of agro-environmental programmes should therefore concentrate on their effects on the environment. EU regulations do not exactly define which production methods are extensive and environmentally favourable and thus deserving of promotion. This definition is largely left to the discretion of the Member States. The guidelines for the regulations only state that Member States have to define the terms 'extensification' or 'extensive production' when considering positive effects on the environment and natural habitats.

Regulation 2078/92 states eight issues that can be supported by the EU:

- substantial reduction in the use of fertilisers and/ or plant protection products as well as introduction or maintenance of organic farming methods
- change of farmland to extensive grassland and keep or introduce more extensive forms of crop including forage production
- reduction of sheep and cattle per forage area
- use of farming practices compatible with the requirements of protection of the environment as well as rearing animals of local, endangered breeds or cultivating threatened agricultural plant varieties
- ensure the upkeep of abandoned farm land or woodlands
- setting aside farm land for at least twenty years for nature protection and the establishment of biotope reserves
- management of land for public access and leisure activities

- improvement of the training of farmers with regard to production methods compatible with the environment.

The regulation demands that farmers voluntarily participate in environmental programmes. However they must commit to the programme for five years. Their obligation must exceed the mere compliance with the principles of 'good agricultural practice' and they must guarantee a considerable reduction of yield-increasing means of production.

As mentioned, the EU contributes 50%, in some cases even 75%, to the financing. The premium is determined by a comparison of gross margins with and without programme participation. Farmers' loss of income (opportunity costs) can be made up by incentive components of a maximum of 20% of the opportunity costs. Not only the introduction of environmentally favourable measures, but also their continuation is remunerable. In this case opportunity costs are determined by incomes from alternative production including all other income components. From the EU's perspective, agro-environmental programmes have proved successful. Therefore from the year 2000 on they will be continued with some modification under a new regulation on promotion of rural areas and a funding increase.

One modification in particular is worth mentioning. Previously the guidelines for the application of Regulation 2078/92 state that 'good agricultural practice' is the starting position of agro-environmental programmes. But from 2000 on, the EU law will stipulate this criterion, thus giving an incentive to increase efforts to put this term in concrete form. From 1993 to 1997 the EU put 1.050 million ECU at Germany's disposal for financing agro-environmental programmes, equivalent to 22% of the overall budget of the programmes. Germany actually spent 918 million ECU, i.e. 87% of the original funding (Niendecker, 1998). In Germany there are, apart from a national programme, more than twenty different agro-environmental programmes at the Länder level. The federal programme promotes mainly extensification measures for arable land and grassland as well as organic production methods. It only refers to the first three of the eight measures in EU Regulation 2078/92. Agro-environmental programmes of the Länder that contain these components are 60% cofinanced by the federal government.

Some agro-environmental programmes have a simple structure and contain only a few components, others are quite comprehensive. The structure of the national programme does not give a realistic impression of the actual priorities of Länder programmes. Meanwhile it has come clear that the trend is toward measures for grassland while there is less emphasis on extensification measures for arable land (Bruckmeier and Langkau, 1996).

Many programmes fail to adequately encourage permanent crops on erosion-prone land. This reflects the greater emphasis placed on extensification further than the protection of nature and development of the cultivated landscape. Promotion of measures concerning land management and public access for leisure activities, instruction, and information for farmers about protection of nature, rearing rare animal breeds, and twenty year set-asides are also rarely applied.

Empty public coffers and the fact that promoting agricultural activities with greater ecological effects costs much money have led to two types of agro-environmental programmes in Germany as well as in the EU. Either they are (a) land-wide but promote only agricultural activities that have minor ecological effects, or (b) they concentrate on agricultural activities with major ecological effects but lack the ambition to involve nearly all farmers. Consequently no Land has developed an agro-environmental programme promoting measures of greater ecological efficiency land-wide.

Examples of the above mentioned types of agro-environmental programmes are those in Baden-Württemberg (type a) and North-Rhine Westphalia (type b). In Baden-Württemberg about 90% of agricultural land is subsidised. The average premium for extensively used land is 130 DM per ha. Extensification of arable land has priority, claiming about 70% of funds, while extensification measures on grassland are supported by about 20% of funds. Nature protection and landscape management receive some 10%. In North-Rhine Westphalia about 3,5 % of agricultural land is subsidised. Premiums amount to approximately 300 DM per ha with an emphasis on measures concerning management of nature and landscape which receive one third of funds. Organic farming is comparatively generously subsidised (30% of funds).

In 1998 a detailed assessment of agro-environmental programmes from an ecological and economic point of view was presented by the Institute of Agricultural Economics of the University of Göttingen with the financial help of the Federal Ministry of Agriculture (Wilhelm, 1999). A

Delphi study was carried out for ecological assessment in which 30 experts in agriculture and landscape ecology were questioned. They were chosen based on their eminent reputation through their publications and/or assessments of environmentally beneficial farming methods. In the Delphi method, experts first give their individual opinions, then all participants are provided with the anonymous results, and finally the experts are asked to give a second group judgement. Through this method it comes to a specific trigger of cognitive processes and and and to an improvement in the quality of the initial judgement. In the Göttingen study, a questionnaire was developed to compare the impacts of different agro-environmental measures on abiotic, biotic, and aesthetic resources. Abiotic resources are water, soil, and atmosphere; biotic resources include the various species of flora, fauna, and biotopes; and aesthetic resources are landscape and cultural aspects. The experts were free to assess agro-environmental measures by differentiating between three regions: regions with potential yields above average, low quality agricultural regions, and low mountain ranges and hillsides. This differentiation was the result of questioning several experts in advance. The experts used a scale from 0 - 5 to assess the impacts of various agro-environmental measures, 5 indicates significant and 0 indicates no environmental improvement when compared to 'good agricultural practice'.

They reported the following: Measures that are notably efficient for promoting protection of abiotic resources are elimination of fertiliser and plant protection products in particularly sensitive areas and those that encourage a change from arable crop production to extensive grassland. Streuobstmeadows (meadows with fruit trees), establishing and managing protected plantations, management agreements for individual areas, and elimination of fertilisers and chemical plant protection are particularly effective for the protection of biotic resources. The establishment of perennial plantings along streams and rivers, protected plantations, woody plants, and Steuobstmeadows are useful tools for protecting aesthetic resources.

The ecological effectiveness of Länder agro-environmental programmes were also evaluated. The programme in Baden-Württemberg is ecologically most effective when the benefits gained by environmental programmes are referred to the total agricultural area. But when we take the agricultural area to which the programmes are actually applied to as a basis, the agro-environmental programme of North-Rhine Westphalia is ecologically most effective.

Agro-environmental programmes were further evaluated in a cost-effectiveness analysis using the results of ecological assessments and the costs of programme application. Agro-environmental programmes that were ecologically most effectively, also spent their financial resources most efficient. The agro-environmental programme of North-Rhine Westphalia is the most expenditure-efficient alternative when costs and efficacy are evaluated for the application area. Baden-Württemberg is most expenditure-efficient among existing programmes when total agricultural area is considered.

It was already mentioned that no Land promotes measures with major ecological efficacy land-wide. It would be economically justifiable for the German Länder to establish such programmes and choose the expenditure-efficient programme of North-Rhine Westphalia as a model. Such change would be a potential Pareto-improvement. Economic benefits arise because of the environmental benefits connected with the provision of agro-environmental goods. Economic costs arise mostly from the excess burden connected with financing the expenditures by taxes. When we equate this excess burden with public expenditures, we can be sure that we are not underestimating the economic costs. In 1996 public expenditures for the agro-environmental programme in North-Rhine Westphalia amounted to some 14.6 million DM. If this programme were applied to 90% of the agricultural area in Germany, public expenditures would amount to about 44 billion DM. Nation-wide, an expenditure-efficient agro-environmental programme, covering 90% of the agricultural area, would thus amount to less than 44 million DM per year. For Germany's 37 million private households, economic costs would be 120 DM per household per year. From the results of already existing analyses of willingness to pay for the maintenance of agricultural landscapes, regional management of landscapes and general preservation of nature and species, we know that the average willingness to pay of private households totals at least 170 DM per year (Hampicke, 1999).

Even without an exact calculation of economic costs and benefits we can state that a new orientation of agro-environmental programmes towards greater ecological effectiveness on larger areas leads to positive welfare effects.

Agriculture and the Environment

Market-orientated Reorganisation of Agro-environmental Policy

This section deals with the question of how to develop the present agro-environmental policy further. From the economic point of view the most important criterion for assessing a policy is whether it contributes to better satisfying the needs of members of a society. As far as providing society with private goods is concerned, the decentralised, error-absorbing, and receptible market system has clearly proved superior to central planning. According to the majority of economists, it is also desirable to profit from the advantages of a market economy when supplying the population with agro-environmental goods. Markets for these collective goods are not as easy to establish as markets for private goods; governmental intervention must be stronger. However, market instruments should be given priority. This strategy can only lead to the desired benefits when certain requirements are fulfilled and essential constitutive criteria are considered. The success of a market orientated remuneration system for ecological activities is determined by the principles of a market system. These are:

Scarcity of services   Not all services of individuals that are appreciated by other individuals have to be allocated by the market. The free market allocation mechanism has only to be applied when demand exceeds supply and an expansion of supply is possible and resource binding.

Price as indicator of scarcity   Consumers, by their willingness to pay, signal potential producers as to which goods shall be produced. The price of goods is determined by the costs of production and their valuation by the consumer.

Conforming to the rules (Ordnungskonformität)   In principle it is clear that suppliers must be compensated for their services. It is also considered acceptable that demanders must pay money to receive services.

Moral behaviour of market participants   Every market participant respects the willingness of other market participants to trade.

Freedom of contract   All members of society are free to decide which markets they wish to act on as suppliers or demanders, and how much they want to offer or to demand.

The first principle states that ecological services of agriculture be in short supply. This requirement is undoubtedly fulfilled. Farmers do not 'automatically' use natural resources in such a way that satisfies society;

they could use resources, albeit at the expense of the production of agricultural goods, in another way.

The second principle stresses the role of demand in a market economy. It is of particular importance for the establishment of a market-orientated remuneration system. The demanders of ecological services from agriculture are interested in the agro-environmental goods provided and not in the activities that led to the provision of them. Therefore, effects on the environment have to be remunerated, not the activities that lead to the desired environmental effects. As far as agro-environmental programmes are concerned, the procedure is different: farmers receive payments for certain activities, e. g. not using pesticides or mowing on a fixed date.

With respect to remunerating activities, programmes are only orientated toward opportunity costs. Thus neither an incentive to provide agro-environmental goods by current production methods as cost-effectively as possible is created, nor the quest for new resource-preserving production methods is encouraged. A remuneration system only has cost-minimising and innovation-stimulating effects when the amount of remuneration is determined by the value of ecological activities to society. This requires the integration of (monetary) demand for ecological services of agriculture in the remuneration system.

Establishment of markets on which the demand for ecological services can be directly realised is restricted as most agro-environmental goods are collective goods. It is, therefore, necessary to use scientific methods of economic assessment of environmental goods to determine the population's willingness to pay. Today there are two categories of valuation methods: methods that assess the monetary value of environmental goods with the help of market information (indirect methods), and methods that do not refer to market information (direct methods).[1] Indirect methods are based on the principles of neo-classical demand theory and on the assumption that the valued nature is important only for people who use nature. Direct methods assume that in experiments or surveys, people act as they would in 'real life'. All valuation

---

[1] For an overview on these methods see Garrod and Willis, 1999.

methods are continuously developed and have an increasingly informative value.

Natural conditions and land use differ widely among regions. The willingness to pay for ecological services must, therefore, be determined for specific regions. This implies that the institutions that evaluate the population's willingness to pay should also be established regionally. The landscape conservation associations existing in some German states (e.g. Bavaria and Saxony) could be models of such institutions. Their task is the coordination and evaluation of conservation measures in cultivated areas and natural biotopes.

When farmers receive the entire willingness to pay, no consumer surplus remains for the demanders. Even when this is irrelevant under allocation aspects, distribution aspects argue for establishing producers' and consumers' surpluses on markets for ecological services. Producer surpluses are essential to induce farmers to treat the land on which they produce agro-environmental goods as a value worth preserving.

According to the principle of conforming to the rules (Ordnungskonformität), it must be determined whether all ecological activities in agriculture shall be remunerated by society, or whether farmers should provide certain agricultural services without compensation. Literature frequently suggests to carry out this examination referring to the concept of negative and positive external effects (Heißenhuber, 1995; Ahrens, 1992; Hofmann, 1995). It is argued that remunerable ecological activities are positive external effects. Costs connected with a reduction of negative external effects are charged to farmers. From the economic point of view, the concepts of negative and positive external effects are unsuitable for defining remunerable ecological activities. Economically, benefits and damage are equal effects, but with reversed premises; a benefit is averted damage, a damage is a missed benefit. Thus, every change in the agricultural use of the environment can be interpreted as a reduction of a negative external effect or an increase in positive external effects. That is, when society wants farmers to reduce ground water pollution, this can be interpreted as a desired reduction of a negative external effect, or as an increase of a positive external effect, i.e. as a contribution toward securing the quality of the water supply.

There is no economic criterion for determining exactly which ecological activities of agriculture are to be remunerated. This becomes clear when interpreting the question of remuneration according to the property right theory (Hampicke, 1996). The individual who sells services for money acts deliberately since he could also retain them. Thus, when an ecological activity of agriculture is remunerated, it is simulated that farmers have the natural resources needed for the activity at their disposal. When defining remunerable ecological services the fundamental issue is the allocation of property rights to natural resources. The determination of property rights is based on value judgements which may change. Thus, the determination undergoes an evolutionary process. There is no social fundamental principle that defines how to proceed; only suggestions can be made about what must be considered. For example, it should be realised that increasing legal uncertainty leads to negative allocation effects. As a consequence, property rights may not be altered due to political opportunity or even arbitrariness. When farmers are denied property rights to natural resources and must provide ecological services for nothing, it must be secured that they can afford it. If economic disadvantages increase to such an extent that farming no longer pays, a small segment of the population would be charged with very high costs to the advantage of the majority of the members of the society. Even the favoured majority would consider such a situation as unjust and therefore unacceptable. The decision as to which ecological services are remunerable and which are not should always be made by binding and understandable regulation with broad public acceptance. In Germany there is consensus that contributions of agriculture to preserve the variety of species and the beauty of the landscape in particular must be remunerated. Many forms of extensive land use which provide these agro-environmental goods (e. g. sheep grazing on oligotrophic grassland), usually cover only a small portion of fixed costs (Hampicke, 1999). Without remuneration, farmers would not permanently provide these activities.

The fourth principle shows that a functioning free market system has moral requirements for individuals. Non-economists have the idea that a free market system leaves no space for morality. They refer to Adam Smith, who in 1776, wrote (Smith, 1776):

> It is in this manner that we obtain from one another the far greater part of those good offices which we stand in need of. It is not from the benevolence of the butcher, the brewer, or the baker, that we expect our dinner, but from their regard to their own interest. We

address ourselves not to their humanity but to their self-love, and never talk to them of our own necessities but of their advantages.

For a correct interpretation of this quotation one has to take into account Smith's views on the principles of individual and social life laid down in his 'Theory of Moral Sentiments' (Smith, 1759). There Smith presents two precepts of individual behaviour: man's desire for prestige and recognition by others, and the desire for self-esteem. Both of these result in correct moral behaviour of individuals. Smith assumes this behaviour when he explains the beneficial qualities of a free market system:

> In the race for wealth, and honours, and preferments, he [every individual; R.M.] may run as hard as he can, and strain every nerve and every muscle, in order to outstrip all his competitors. But if he should justle, or throw down any of them, the indulgence of the spectators is entirely at an end. It is a violation of fair play, which they cannot admit of.

One of the most important rules of a market economy is that market participants respect the willingness to exchange of those with whom they conduct market transactions. Accepting others' willingness to exchange can not be explained only by farsighted behaviour of individuals or a nearly perfect control system. Acceptance also requires moral behaviour of individuals, i.e. their willingness to observe laws for the laws' sake. Expressed in economic terms, market participants show their willingness to pay for the non-violation of norms because they would rather renounce market goods or money than disrespect other market participants' willingness to exchange. The significance of moral behaviour of market participants is proved by numerous economic-historic studies on the establishment and development of market economies and econometric studies on the dependence of individual morals and economic growth (compare e.g. Knack and Keefer, 1997; Kubon-Gilke, 1997, p. 246-248). Market economies cannot work without individual morality, without deliberate observance of laws; markets are not moral-free zones. On the contrary, individual morals play an important role in making free-market coordination superior to other coordination mechanisms.

Consequences of this market principle for establishing a remuneration system arise in connection with the last principle, the freedom of contract. According to this principle, every member of society is free to choose to act on a market. A remuneration of agricultural ecological services that takes this principle into account does not determine for farmers the extent to which they will provide agro-environmental goods.

The current, widely accepted principle of sustainable development requires that the use of natural resources must also consider the interests of future generations. Of course, there is no certainty about their preferences. However, we can assume that the interests of following generations, as far as physical needs are concerned, will not differ substantially from interests of today's generations. Also future generations will not be prepared to accept environmental deterioration that will endanger the fulfilment of their needs. Hampicke (1992, p. 314f.) explains in detail ecological conditions that must be fulfilled to meet existential human needs in the long run. These conditions can be summarised as an ecological goal with three components: preservation of the variety of species and biotopes, securement of the self-regulating ability of ecosystems, and the maintenance of a specified state of several large biogeochemical systems.

It was previously mentioned that ecological activities of agriculture in the area of biodiversity must be remunerated. It follows from the principle of sustainable development that these activities must be divided in two types: ecological activities that preserve endangered species, and ecological activities for unendangered species. The principle of freedom of contract can unrestrictedly be applied to the second type of activities. The remuneration system for the first type of ecological activities may consider this principle only when society's willingness to pay is so high that it exceeds the costs for preservation of the variety of species. The stronger the market's ethics, the sooner this condition will be fulfilled. The definition of morality in markets makes whether the parties know each other or not insignificant. The willingness of others to exchange has to be accepted, regardless of whether these persons are identifiable business partners or anonymous market participants; homo oeconomicus must also carry out his duty toward strangers. Future generations are 'strangers' for the current market participant, as are all market partners at the end of the chain of market transactions, and whose interests he must consider when making decisions. Therefore, a close relationship exists between the ethics of the market and ethics of the future that includes individuals' respect for the coming generation's willingness to exchange agro-environmental goods. We only have to assume that future generations will not have no lower ethical values than people living today. An ethical difference between current and future generations is sometimes explained by the purely hypothetical existence of future generations. This deliberation is meaningless within the framework of market ethics, as market ethics demand that any acting person also considers the interests of people living far away and whom he will never meet. These are also hypothetical persons for the individual, are people who will not be born for a hundred years.

As mentioned, respecting the willingness of future generations to exchange demands the preservation of biodiversity. When people behave ethically correctly in the market, they also have a willingness to pay to secure the variety of species. Whether this non-use value together with the use value is high enough to secure the preservation of the variety of species in a market-conforming system can only be clarified by empirical studies. The present studies on costs and benefits of natural conservancy all arrive at the same result: in Germany the willingness to pay for the preservation of nature exceeds not only its economic costs, but also its financial costs (Geisendorf et al., 1999; Hampicke, 1999).

Summing up the discussion we can say: the remuneration of ecological activities of agriculture conforms to market principles when:
- results, not actions, are remunerated
- the amount of remuneration is determined by the value, not the cost, to the public of activities
- regional markets are established
- producer and consumer surpluses are created
- it focusses on ecological activities in the field of biodiversity

Of course the observance of these requirements is not sufficient to ensure that the remuneration system will lead to the desired benefits. Market-conforming remuneration of ecological activities of agriculture is the application of the market idea to a completely new field. Theoretical consideration cannot foresee implementation problems. Therefore such a system must be carefully applied, initially in large-scale trials to gain experience. The advantages of a functioning, market-conforming remuneration system should be worth such trials.

References

Ahrens, A. (1992). Gesellschaftliche Aspekte der Honorierung von Umweltleistungen der Landwirtschaft. In Untersuchung zur Definition und Quantifizierung landespflegerischer Leistungen der Landwirtschaft nach ökologischen und ökonomischen Kriterien und ihre Umsetzung in Umweltberatung und Agrarpolitik. Vorstudie im Auftrag des Bayerischen Staatsministeriums für Landesentwicklung und Umweltfragen. Weihenstephan.

Baldoch, D. and Lowe D. (1996). The Development of European Agro-Environment Policy. In M. Whitby (ed.), The European Environment and CAP Reform Policies and Prospects for Conservation. Wallingford: 8-25.

Bauer, S. und Thielcke, G. (1982). Gefährdete Brutvogelarten in der Bundesrepublik Deutschland und im Land Berlin: Bestandsentwicklung, Gefährdungsursachen und Schutzmaßnahmen. In Vogelwarte 31: 183-391.

Beck, R. (1996). Die Abschaffung der „Wildnis": Landschaftsästhetik, bäuerliche Wirtschaft und Ökologie zu Beginn der Moderne. In Konold, W. (Hrsg.): Naturlandschaft – Kulturlandschaft: die Veränderung der Landschaften nach der Nutzbarmachung durch den Menschen, 27-44, Landsberg.

Blab, J. (1992). Isolierte Schutzgebiete, vernetzte Systeme, flächendeckender Naturschutz? – Stellenwert, Möglichkeiten und Probleme verschiedener Naturschutzstrategien. In Natur und Landschaft 67/9: 419-424.

Blab, J. und Kudrna, O. (1982). Hilfsprogramme für Schmetterlinge. Ökologie und Schutz von Tagfaltern und Widderchen. Greven: Naturschutz aktuell Nr. 6.

Brown jr., G. M. (1993). Economics of Natural Resource Damages Assessment; A Critique. In J. Kopp and K. Smith (eds.), Valuing Natural Assets. Washington D. C.: Resources for the Future 73-105.

Bruckmeier, K. und Langkau, J. (1994). Eine grüne GAP? Ökologisierung der Europäischen Agrarpolitik. Köln: Katalyse-Institut für angewandte Umweltforschung.

Garrod, G. and Willis, K. (1999). Economic Valuation of the Environment – Methods and Case Studies. Cheltenham, Northampton: Edward Elgar.

Geisendorf, S. et al. (1999). Die Bedeutung des Naturvermögens und der Biodiversität für eine nachhaltige Wirtschaftsweise. Berichte des Umweltbundesamtes 6/98. Berlin: Erich Schmidt Verlag.

Haber, W. (1972). Grundzüge einer ökologischen Theorie der Landnutzungsplanung. In „Innere Kolonisation" 21: 294-298.

Haber, W. (1985). Rahmenbedingungen der Landwirtschaft aus der Sicht der Umwelterhaltung. In Bayerisches Landwirtschaftliches Jahrbuch 62, Sonderheft 1/1985: 15-23.

Haber, W. (1988). Von extensiver zu intensiver Nutzung – Wandlungen in der agrarischen Landnutzung aus ökologischer Sicht. In Schriftenreihe Deutscher Rat für Landespflege 54: 265-268.

Hampicke, U. (1992). Ökologische Ökonomie. Teil 4: Individuum und Natur in der Neoklassik, Natur in der ökonomischen Theorie. Opladen: Westdeutscher Verlag.

Hampicke, U. (1996). Perspektiven umweltökonomischer Instrumente in der Forstwirtschaft insbesondere zur Honorierung ökologischer Leistungen. Stuttgart: Metzler-Poeschel.

Hampicke, U. (1999). Conservation in Germany's Agrarian Countryside and the World Economy. In A. K. Dragun, C. Tisdell (eds.), Sustainable Agriculture and Environment. Globalisation and the Impact of Trade Liberalisation. Cheltenham: Edward Elgar 135-152.

Heißenhuber, A. (1995). Betriebswirtschaftliche Aspekte der Honorierung von Umweltleistungen der Landwirtschaft. *Agrarspectrum* 24: 123-141.

Höll, H. and Meyer, H. v. (1996). Germany. In M. Whitby (ed.), The European Environment and CAP Reform Policies and Prospects for Conservation. Wallingford: CAB International 70-85.

Hötzel, H.-J. (1996). Umweltvorschriften für die Landwirtschaft. Stuttgart: Eugen Ulmer.

Hofmann, H. (1995). Umweltleistungen der Landwirtschaft – Konzepte zur Honorierung. Stuttgart, Leipzig: Teubner Verlagsgesellschaft.

Kemper, H. (1992). Erfahrungen mit dem Kooperationsmodell aus Sicht der Landwirtschaft. In Rheinisch-Westfälisches Institut für Wasserchemie und Wassertechnologie GmbH (ed.), Erfahrungsaustausch über regionale Ergebnisse der Kooperation Wasserwirtschaft/Landwirtschaft in Nordrhein-Westfalen. Duisburg: 71-83.

Knack, S. and Keefer, P. (1997). Does Social Capital have an Economic Payoff? A Cross-Country Investigation. *Quarterly Journal of Economics*: 1251-1288.

Korneck, D. und Sukopp, H. (1988). Rote Liste der in der Bundesrepublik Deutschland ausgestorbenen, verschollenen und gefährdeten Farn- und Blütenpflanzen und ihre Auswertung für den Arten- und Biotopschutz. In Schriftenreihe Vegetationskunde 19, Bonn-Bad Godesberg.

Kubon-Gilke, G. (1997). Verhaltensbindung und die Evolution ökonomischer Institutionen. Marburg: Metropolis-Verlag.

Niendiecker, V. (1998). Die Ratsverordnung (EWG) Nr. 2078/92 als Instrument der europäischen und nationalen Agrarumwelt- und Agrarstrukturpolitik. *Berichte über Landwirtschaft* 70: 520-539.

OECD (1994). Managing the Environment: The Role of Economic Instruments. Paris.

Osterburg, B., Wilhelm, J. and Nieberg, H. (1997). Analyse der regionalen Inanspruchnahme von Agrarumweltmaßnahmen gemäß Verordnung (EWG) 2078/92 in Deutschland, Arbeitsbericht 8/97. Braunschweig: Institut für Betriebswirtschaft der FAL.

Remmert, H. (1988). Naturschutz: ein Lesebuch nicht nur für Planer, Politiker und Polizisten, Publizisten und Juristen. Berlin, New York, London, Paris, Tokyo: Springer-Verlag.

Schumacher, W. (1997). Naturschutz in agrarisch geprägten Landschaften. In Erdmann, K.-H. und Spandau, L. (Hrsg.): Naturschutz in Deutschland: Strategien, Lösungen, Perspektiven, Stuttgart: Thiene, 95-122.

Smith, A. (1759). The Theory of Moral Sentiments, 1976 edition. A. L. Macfie and D. D. Raphael (eds.), Glasgow Edition of the Works and Correspondance of Adam Smith. Oxford: Clarendon Press.

Smith, A. (1776). An Inquiry into the Nature and Causes of the Wealth of Nations, 1976 edition. R.H. Campbell et al. (eds.), Glasgow Edition of the Works and Correspondance of Adam Smith. Oxford: Clarendon Press.

Völsch, W. (1993). Entschädigungs- und Ausgleichsrecht in den Wassergesetzen - Rechtsgrundlage und praktische Auswirkungen, insbesondere auf landwirtschaftliche Belange. Köln.

Wilhelm, J. (1999). Ökologische und ökonomische Bewertung von Agrarumweltprogrammen. Delphi-Studie, Kosten-Wirksamkeits-Analyse und Nutzen-Kosten-Betrachtung. Frankfurt am Main: Peter Lang.

Winkler, W. (1994). Ordnungsgemäße Landwirtschaft und Umweltrecht. In R. Breuer et al. (eds.), Jahrbuch des Umwelt- und Technikrechts 1994. Heidelberg: Deckers und Schenk, 545-586.

Chapter 12
Research and Technology in German Agriculture

Joachim von Braun and Matin Qaim
Center for Development Research (ZEF), University of Bonn

Introduction

While the expansion of cultivated area had been the major source of worldwide production gains throughout agricultural history, in the last century scientific advances more and more led to a substitution of human knowledge for natural resources. Today, most of the global increases in food and fiber production originate from technological progress and growth in the productivity per unit area. Technological progress is research driven. Stirred by the need to feed their populations and wishing to take advantage of the scientific revolution, by the middle of the twentieth century almost all countries had instituted a national agricultural research system (NARS). This chapter describes and analyzes the German NARS, one of the oldest worldwide. The initial patterns of institutionalizing agricultural research in Germany were later also embraced as a model for the establishment of research systems in a number of other industrialized countries (Ruttan 1982).

The present situation of the German NARS must be viewed in a dynamic fashion, taking into account both the past and the future. The chapter starts with a brief historical overview before discussing the changing national and international framework of agricultural research and development (R&D). Although the relative importance of a narrowly defined agricultural sector is declining in Western Europe, the demand for agricultural research remains high. The research agenda has broadened due to new emerging requirements and changing preferences of the population. In the next section the structure of public agricultural research in Germany is presented. Studies of various aspects of the German NARS have been mainly confined to research in the public domain (e.g. Burian 1992; Koester and von Cramon-Taubadel 1994; Brinkmeyer 1996; Hockmann and Recke 1997; Isermeyer and Werner 1997). The private sector, however, accounts for a substantial proportion of the overall agricultural research activities in Germany and its importance is increasing due to globalization tendencies and an international strengthening of intellectual property rights (IPRs). Consequently, a special section gives an overview of private agricultural R&D activities and investments in Germany. After that Germany's contribution to international agricultural research is considered. Some suggestions for system restructuring and some wider conclusions for agricultural research policy in industrialized countries are discussed in the final section.

Historical Background and Changing Framework Conditions

The roots of the German agricultural research system date back to the early 1800s. The efforts at that time were among the first worldwide to institutionalize agricultural science (cf. Ruttan 1982). Albrecht Thaer, a medical doctor in Berlin, initiated the establishment of independent agricultural academies in all German states during the first half of the nineteenth century. However, these academies were primarily teaching institutions. It was Justus von Liebig, with his pioneering work on agricultural chemistry published in 1840, who initiated agricultural research based on the natural sciences. In the following decades, a great number of publicly supported agricultural experiment stations were founded in the different states of Germany. The aim was to create a link between science and agricultural practice. Correspondingly, agricultural faculties were established in several universities with combined educational and research objectives, and existing agricultural academies extended their mandates to also include research. Following the evolution of Germany's political structure, a more centralized approach to agricultural research was begun in the early twentieth century with the foundation of federal research centers. These federal centers partly replaced the state-level experiment stations (Reichrath 1990). Most of the independent academies were integrated into the university system during the National Socialist period. After World War II, the Federal Ministry of Agriculture (BML) took on the primary responsibility for nonuniversity public agricultural research, whereby existing federal research centers as well as newly created ones became the executing entities. Today, there exists a network of ten federal agricultural research centers, each with a specified mandate (BML 1999). Whereas the federal centers predominantly have an applied focus, the university faculties, which are administered at the state level, often carry out more basic agricultural research. Since Germany's reunification, there are agricultural faculties at ten different universities, plus institutes with agriculture-, food- or forestry-related areas at several other universities and technical colleges.[1] A more comprehensive overview of the present research system structure and the role of different research actors in the public domain is given in the next section.

Although the chief aim of agricultural research, the augmentation of agricultural productivity, is still topical in Germany, the conditions or the political, economic, social, environmental and technological framework changed substantially over time. Historically, the main driving force for research had been the desire to be able to produce enough food while shifting labor from the primary sector to industry. With rising overall living standards, the objective of increasing farm incomes gained in importance. However, national and international developments during the last twenty years brought about completely new challenges for agricultural research and led to a broadening and fine tuning of research priorities. Some of the main recent developments that affect European and German agricultural research are discussed in the following (also see Paillotin 1998). While the various new aspects are listed separately in reality they are often closely interlinked.

Globalization

The Uruguay Round of the General Agreement on Tariffs and Trade (GATT) for the first time explicitly included liberalization of agricultural trade in the talks. Although the barriers to international trade are still

---

[1] The term technical colleges as used here refers to one particular element of the German university system, the universities of applied science ("Fachhochschulen").

considerable, the trend is clear: domestic producers are gradually being confronted with competition from abroad. Apart from highlighting necessary structural adjustments in the agricultural sector, it is also the task of domestic research to provide the technological tools needed for achieving and sustaining international competitiveness. Moreover, in the developing world the demand for food imports is growing and will have to be met by exports from industrialized countries at reasonable world market prices. Yet globalization also comprises a liberalized flow of knowledge and technology. Hence, not only the agricultural sector itself but also the German NARS, the sector of knowledge production, is challenged to embrace the concept of international comparative advantages.

Agricultural Policy Reform in the European Union (EU)

Partly driven by obligations of the GATT and the World Trade Organization (WTO), but also by the wish to reduce public expenditure, in recent years the Common Agricultural Policy (CAP) of the EU has experienced various reform steps. The reduction of trade barriers and of administrative regulations gradually leads to closer world market orientation. In the longer run, this will improve the efficiency of agricultural production. In the short to medium run however, a greater demand for disaggregated policy analysis arises, so that the reform processes can be managed in a socially acceptable way.

German Reunification

The German reunification in 1990 also changed the conditions of agricultural research. The structure of the agricultural sector in the former German Democratic Republic (GDR) is completely different from the one in Western Germany. This has to be accounted for in the formulation of research objectives. In particular, efforts to privatize the former collective or state-owned agricultural enterprises brought new topics to the research agenda of social scientists. However, reunification not only affected the contents of research. Much more fundamental are the institutional implications engendered by the merger of the two innovation systems. To achieve an appropriate size of the new system, the justification of all existing institutes had to be scrutinized. Some of the academic institutes were closed down, others were transformed or integrated, and some new institutes emerged. Unsurprisingly, German agricultural research investments showed a noteworthy increase after 1990 compared with the investments in West Germany before the reunification. The process of system restructuring has not yet been completed and gradual budget cuts are scheduled for the future.

Economic Transition in Eastern Europe

The collapse of the communist regimes and the economic transition in Eastern Europe also had ramifications for agricultural research in Germany. The economic slump and the worsening agricultural situation in many of these countries call for agronomic as well as policy-oriented research. These tasks can hardly be fulfilled by the national knowledge systems of Eastern Europe alone, as they are currently rather underdeveloped, even deteriorating (cf. Csaki 1998). Western European NARS are challenged to bridge the research gap for partnership considerations, but also with a view to the upcoming EU eastward enlargement.

Focus on Sustainability

During the last two decades, knowledge about interlinkages between human actions and ecological systems has increased remarkably. Hence, attention to sustainable utilization of natural resources, as called for in Agenda 21 of the Rio conference in 1992, is growing. Concerns are particularly pronounced in Europe, because the effective demand for environmental quality rises with increasing living standards. In Germany, where agricultural production intensities are high, negative environmental externalities of farm technologies have become a serious political issue. The growing awareness of environmental problems extended the scope of existing agricultural research programs and introduced a substantial number of new topics to the R&D agenda.

Changed Consumer Preferences

The fears of acute food shortages have disappeared in Western Europe, but consumer demand for food quality is rising. In addition to quality attributes related to the end product, aspects of the production process increasingly influence consumer attitudes. Hence, issues such as environmental side effects and animal welfare must also be considered as quality components. The altered demand patterns require agricultural and food scientists to develop refined production and processing technologies. Furthermore, comprehensive knowledge about the multifaceted determinants of consumption decisions becomes an important ingredient in satisfying food preferences and expectations of the German population.

Technological Innovations

During the last twenty years, scientific developments effected breathtaking technological progress. This has an impact not only on agricultural production, but also expands the research horizon. The speed at which data can be processed and transmitted through advances in information and communication technologies is a case in point. Somewhat more closely related to agriculture is biotechnology. New techniques of identifying, isolating, conserving and transferring biological material augment the economic value of genetic resources (cf. Virchow 1999). The use of biotechnology in plant breeding will facilitate the accomplishment of completely new crop traits (e.g. the production of pharmaceuticals or biodegradable plastics within the plant). This could revolutionize the role of agriculture (Abelson and Hines 1999). The responsible management of such technological developments requires new interdisciplinary research approaches. As the current debate on genetically modified crops in Europe demonstrates, it is not enough to develop new technologies. Agricultural scientists also need to become actively engaged in communicating such innovations to the broader public in a trustworthy way.

Increased Role of Private Sector R&D

Although private firms in industrialized countries always made up a significant share of overall applied agricultural R&D, in recent times different developments led to a growing role for the private sector (e.g. Byerlee 1998). Globalization brings about an international strengthening of IPRs and opens up new opportunities for transboundary technology transfers. Furthermore, biotechnology contributes to the trend of chemical companies becoming transnational life sciences firms making huge R&D investments, some in areas of basic biological research. Given the limited availability of public funds, agricultural research policy is challenged to more explicitly identify the comparative advantages of public research and to develop models and incentives for efficient cooperation between the public and the private sectors.

The mentioned developments and other factors have major implications for the direction and scope of agricultural research in Europe and in Germany. The demand for agricultural research remains high, but continuous institutional adjustment processes within the NARSs are necessary to efficiently satisfy this demand. It is important to stress, however, that there is more than a unidirectional relationship between the economic context and agricultural research; inversely, agricultural research must also be seen as a powerful determinant shaping the global economic context (cf. von Braun 1998; von Braun and Wehrheim 1995).

Public Research

The development of public agricultural research expenditures in Germany during the last two decades is shown in Figure 1, differentiated between university and nonuniversity research. In nominal terms, the R&D expenditure doubled from 1981 to 1991. As the earlier figure only refers to Western Germany, part of this effect is explained by the German reunification. But also the figures for the 1990s clearly indicate a rising tendency.

A meaningful method to compare the level of agricultural research expenditures across countries is to express them as a percentage of the corresponding agricultural gross domestic product (GDP). The resulting ratio has been termed agricultural research intensity (ARI) by Anderson et al. (1994). The ARI is also shown in Figure 1.

The upward trend of the ARI is not surprising since the agricultural GDP did not significantly change over the period of consideration. In an international comparison, the German figure of around 5 percent is rather high, matched only by the Scandinavian countries and Canada in the early 1990s. For comparison: developing countries usually have an ARI of less than 1 percent; the USA has an intensity of 2.5 percent (data provided by ISNAR). Yet such an international comparison is difficult, because data sources are not always consistent across countries.

Figure 1:
Public agricultural research expenditures in Germany (1981, and 1991–1995)

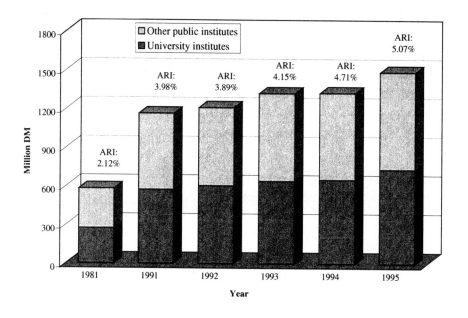

Note: ARI is the agricultural research intensity, i.e. the research expenditure relative to the agricultural GDP.
Sources: Based on BMBF (1998) and BML (various issues).

Another interesting indicator is the relationship of agricultural research expenditure to total public research expenditure. For Germany, this ratio was 4.2 percent in 1996 (BMBF 1998). This appears high on first sight, because the agricultural sector only contributes 1.2 percent to the total national GDP and 3.5 percent to the overall labor force. Defining the role of agriculture in such a narrow sense is, however, misleading because there are substantial backward and forward intersectoral linkages that must not be neglected. Furthermore, in Germany agriculture, excluding forestry, still covers some 55 percent of the total land area and food consumption expenditures make up about 16 percent of the average household budget. Consideration of the broadened agricultural research agenda, particularly in terms of food quality and sustainable land use management, creates a new perspective for the interpretation of an appropriate food and agricultural research expenditure ratio.

Figure 2:
Main public agricultural research-funding and -implementing organizations

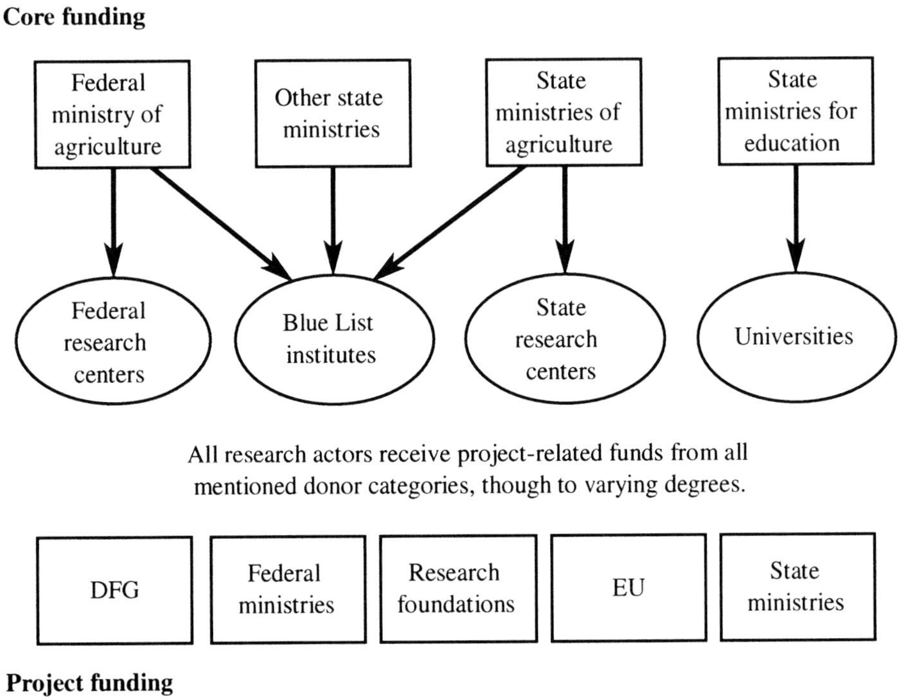

Core funding

```
Federal ministry of agriculture    Other state ministries    State ministries of agriculture    State ministries for education

Federal research centers    Blue List institutes    State research centers    Universities
```

All research actors receive project-related funds from all mentioned donor categories, though to varying degrees.

```
DFG    Federal ministries    Research foundations    EU    State ministries
```

Project funding

Human and Financial Resources

In the mid-1990s, Germany had about 5,000 agricultural scientists working in public research organizations. The allocation to different research topics was as follows: 31 percent were working on plant aspects, 20 percent in forestry, 15 percent on topics of animal production, 14 percent in economics, 11 percent on soil issues and 9 percent on questions of agricultural machinery and technical equipment (Isermeyer and Werner 1997). The pattern of research-funding and implementing organizations is graphically shown in Figure 2. The following paragraphs briefly describe the mandate and structure of the different organizational categories.

Main Research Actors

Universities

Most of the resources available for public agricultural research in Germany are tied to the universities, which also have teaching obligations apart from their research mandate. There are agricultural faculties at ten different universities (Berlin, Bonn, Giessen, Göttingen, Halle, Hohenheim, Kassel, Kiel, Munich and Rostock) in addition to various faculties of horticulture, forestry and food technology. Moreover, the technical colleges ("Fachhochschulen") with agricultural institutes carry out some applied research. The core funding for the universities is decentralized, coming from the responsible state ministries for education. Usually, university professors have the freedom to choose the research topics to be carried out in their departments so that both basic and applied components are covered according to individual preferences. Given the recent cutbacks in institutional budgets, however, direct project support is gaining importance. Such project-related third party resources have to be acquired from external funding organizations through a competitive review process. In 1993, universities raised about 56 million DM from external sources, some 8 percent of their total research budget (Henze 1997). The main third party funding organizations for agricultural research are the German Research Society (DFG), the Federal Ministry for Education and Research (BMBF), the federal and state ministries of agriculture, the EU, various research foundations and private industry (cf. Müller and Brinkmeyer 1996). If the current university research capacity is to be sustained, the need to raise funds externally will grow because core funding is partly allocated according to student numbers and the number of students enrolled in the agricultural faculties is shrinking.

Federal Research Centers

The ten federal research centers form the second largest category of public agricultural research organizations in Germany. They are directly subordinate to the Federal Ministry of Agriculture (BML), and are mandated to carry out applied research to support decision-making processes in the formulation and implementation of agricultural policies. The largest entities, each with several institutes, are the Federal Agricultural Research Center Braunschweig (FAL), the Federal Biological Research Center for Agriculture and Forestry (BBA) and the Federal Center for Breeding Research on Cultivated Plants (BAZ). Although the scientists enjoy some degree of independence in the choice of research topics and the choice of methodologies, departmental research by nature is less autonomous than university research. Moreover, the federal research centers also carry out sovereign duties and tasks of long-term monitoring, that are important for government work but contribute little to the scientific reputation of the researchers involved (Lückemeyer 1999). To guarantee continuity and to compensate for possible scientific disincentives, most of the researchers in the federal centers have secure and permanent employment contracts. Departmental research is not subject to systematic scientific evaluation and third party funding is comparatively rare. The allocation of BML funds to the system of federal research centers has been significantly extended after the German reunification, but a 30 percent reduction in financial and human resources is planned by the year 2005 (BML 1998).

State Research Centers

Similar to the structure at the federal level, there are also research centers at the state level carrying out departmental tasks on behalf of the state ministries of agriculture. Most of the work done in these centers has a very applied focus, including experimental field tests and sovereign duties pertaining to regulations of plant and animal breeding. The state research centers often have close links to farms and thus they also assume extension services. There are at least twenty-five such research centers in the different states of Germany plus a large number of smaller experiment stations.

Blue List Institutes

The so-called Blue List institutes are research organizations cofunded by the federal and state governments. The name 'Blue List' came from the list of these institutions which was printed on blue-colored paper. In total, there are around eighty Blue List institutes, seven of which have an agricultural focus; five of the seven were founded after the reunification, partly replacing academic organizations of the former GDR. The largest ones are the Research Institute for Animal Biology (FBN) and the Center for Agricultural Landscape and Land Use Research (ZALF). In many cases, the agriculture-related Blue List institutes specifically analyze environmental linkages, but there is also one that looks into economic aspects of Agricultural Development in Central and Eastern Europe (IAMO). The Federal Ministry of Agriculture (BML) is the responsible authority at the federal level; yet Blue List institutes enjoy much greater independence than do departmental research centers. As for the universities, the level of third party funding is increasing, though still comparatively low.

Other

In addition to the mentioned university and nonuniversity institutes, there are also other independent public organizations carrying out advanced research related to agriculture. These organizations are usually cosponsored by federal and state governments, but specific project funding, some from the private sector, is also common. Noteworthy is the Max-Planck-Institute for Breeding Research with a clear focus on plant molecular genetics. Some of the Helmholtz-Research-Centers also perform research relevant to agriculture.[2] However, since these organizations often conduct rather basic research, a clear-cut division between agriculture and nonagriculture is not always possible.

---

[2] The Hermann-von-Helmholtz-Society embraces sixteen public research centers which carry out strategic research of government interest in various disciplines. These centers are usually well equipped and enjoy a fairly high degree of scientific autonomy.

Table 1 gives an overview of how the public sector's financial and human resources were split among the main research actors in 1995. Of the almost 1.5 billion DM annually invested, half was spent by the universities (including technical colleges), and the other half by the different nonuniversity organizations. About 58 percent of the agricultural scientists are working in the universities, so that the ratio of financial resources per researcher is lower than in most nonuniversity organizations. This holds particularly when compared with the federal research centers.

Table 1:
Public agricultural research indicators by organizational category (1995)

Category	Number of full-time equivalent researchers	R&D expenditure (million DM)	Expenditure per Researcher (million DM)
Universities [a]	2,970	733	0.247
Federal research centers [b]	834	415	0.498
State research centers [c]	950	199	0.209
Blue List institutes [b]	262	97	0.370
Other [d]	139	42	0.302
Total	5,155	1,486	0.288

[a] Including research activities of the technical colleges.
[b] 1997 figures.
[c] Extrapolated estimates based on figures referring to West Germany before the reunification.
[d] Among other organizations, this includes the Max-Planck-Institutes and the Helmholtz-Research-Centers.
Sources: Based on BMBF (1998), BML (1998), Isermeyer and Werner (1997) and Brinkmeyer (1996).

Agricultural Economics

The German agricultural economics community embraces around seven hundred researchers engaged at university and nonuniversity institutes (Isermeyer and Werner 1997). While about 20 percent of all agricultural researchers in the university faculties are economists, the share of agricultural economists in nonuniversity research is only about 5 percent.

The comparatively low share of economic topics in the departmental research centers is a bit surprising, but ministries also draw on university-based agricultural economics research. The overall community is organized in the German Association of Agricultural Economists (Gesellschaft für Wirtschafts- und Sozialwissenschaften des Landbaues - GEWISOLA), that celebrated its fortieth anniversary in 1999. GEWISOLA convenes annual academic conferences and the conference proceedings constitute an important documentation of German agricultural economics research.[3]

Table 2:
Topics of the annual GEWISOLA conferences (1990-1999)

Year	Conference topic
1990	The Agro-Food Sector within the Single European Market and the International Division of Labor
1991	International Agricultural Policy and the Development of the World Agricultural Economy
1992	Structural Adjustments of the Agro-Food Sector in Central and Eastern Europe
1993	Agriculture and Societal Demands
1994	Agriculture in the European Union after the Common Agricultural Policy Reform
1995	Agricultural Policy and Structural Development of Rural Sectors
1996	World Agricultural and Food Markets under Changing Framework Conditions
1997	The Agro-Food Sector in an Enlarged European Union
1998	The Agricultural Sector in the Modern Information Society
1999	Competitiveness and Entrepreneurship in the Agro-Food Sector

Source: GEWISOLA (various issues).

In the 1950s, agricultural market development and farm management topics dominated the research agenda of German agricultural economists.

---

[3] Apart from the GEWISOLA conference proceedings there are three journals published in Germany with agricultural economics contributions ('Agrarwirtschaft', 'Berichte über Landwirtschaft' and 'Quarterly Journal of International Agriculture', the latter published in English).

While the analytical tools continued to be refined, the focus on basic farm management aspects remained a priority in the GDR (Rosenkranz 1999). In West Germany, however, sophisticated quantitative methodologies of market and policy analysis gained popularity during the 1960s (Reisch 1999). Agriculture was one of the first sectors with a common policy under the European Economic Community. Thus, early on agricultural economists started analyzing the economics of integration and the agricultural markets at the European level. A distinct expansion of the focus to include international markets and agricultural sectors in countries beyond Europe can be observed since the early 1970s. During the last two decades, the research schedule was broadened substantially (see previous section). The topics of the last ten annual GEWISOLA conferences are listed in Table 2. They give an indication of research priorities in the German agricultural economics community.

In general, agricultural economics research in Germany is fairly international, with close links to research institutions in neighboring countries and North America. For the future, it is expected that such international links will become even closer through the increasing export orientation of European agricultural and food industries.

Constraints in the Public Research System

The present structure of public agricultural research in Germany is the outcome of various historical developments. The federal political structure, with quite some degree of independence at the state level, is the main reason for the great number of different research organizations. The fragmentation of the system has been further intensified through the creation of new institutes after the German reunification. As was already argued, the demand for agricultural research is high and even increased in recent years due to a significantly broadened research agenda. However, given the changing framework conditions, it is important to question whether the present organizational structure of the NARS will be able to satisfy this demand efficiently. Whereas in the past, agricultural problems often had a rather location-specific dimension, in many cases globalization tends to diminish the relevance of geographic boundaries in the setting of research priorities. The result is that the same or very similar research topics are carried out in different institutes with lack of coordination. Although informal connections between individual scientists exist, the system provides too few incentives for more formal interorganizational linkages (cf. von Braun 1999). This holds true for both university and nonuniversity institutes. International competition in agricultural research is increasing, and Germany can only establish itself on the global market for knowledge production when the available resources are efficiently pooled. More concrete suggestions for system restructuring are discussed in the concluding section.

Private Research

For the last fifty years, private agricultural research investments were important in industrialized countries, and the contribution of the private sector to overall agricultural research is further increasing (cf. Byerlee 1998). In Germany in particular, private agricultural research is a long standing tradition. Many German firms have established themselves as successful global players in the international market. This means that private research carried out in Germany creates significant transboundary spillovers of technologies embodied in exported agricultural inputs.

Obtaining exact data on private agricultural research investments is difficult. Large, transnational enterprises publish their R&D outlays, but they are often carrying out research in different branches (e.g. agrochemicals and pharmaceuticals). Attributing a certain expenditure amount to the agricultural sector is sometimes difficult. Smaller and more specific firms, on the other hand, are hesitant to disclose information about their R&D activities. Instead of collecting data from individual firms, we therefore decided to analyze R&D activities at the level of the main industry branches for agricultural inputs. Average estimates of the industries' R&D ratios (i.e. R&D expenditure relative to the industry turnover) were obtained from industry associations. The figures were cross-checked with representatives of various private firms. Knowing the industries' turnovers, these percentage estimates can be translated into absolute expenditure amounts. The resulting data are shown in Table 3. Unfortunately, uniform turnover data were not available. In some instances, the stated figures refer to the input sales in the domestic market instead of representing the global turnover of German firms. Given that the value of exports of agricultural inputs is higher than the value of imports, this leads to an underestimation of the German private sector R&D activities in the respective industries. Another drawback is that the area of private animal breeding has been disregarded. As it is mostly carried out at the individual farm level, aggregate data are not easily at hand. Notwithstanding these inaccuracies, an impression of the dimension of private agricultural R&D investments in Germany can be gained.

Table 3:

Estimated R&D expenditures of German firms in different agricultural input industries (1997)

Industry	Turnover (million DM)	R&D ratio (percent)	R&D expenditure (million DM)
Plant breeding	1,410[a]	16.2	228[d]
Crop protection	5,940[b]	10.0	594
Fertilizer	2,860[b]	0.9	26
Animal health [c]	1,360[a]	8.0	109
Animal feed	8,580[a]	0.2	13
Machinery	7,200[b]	5.0	360
Total	27,350	4.9	1,330

[a] *Refers to the domestic German market.*
[b] *Refers to the production of German firms, for both the domestic market and export sales.*
[c] *The animal health industry includes the market for feed additives.*
[d] *This figure has been derived within a detailed firm survey carried out by BDP (1999).*
Sources: BDP (1999), IVA (1998), LAV (1998), FEDESA (1997), supplemented by personal contacts with representatives from various firms and industry associations.

Comparing agricultural research expenditures of the private sector with those of the public sector is difficult because the quality of data is not uniform. Nonetheless, we aggregate the private data for 1997 and the public data for 1995 (see Table 1), obtaining a total annual agricultural research expenditure in Germany of 2.8 billion DM. Of this total amount, the private sector accounts for 47 percent, approximately the private sector contribution in other industrialized countries in the early 1990s: 54 percent in the USA, 51 percent in Japan, 53 percent in France (cf. Alston et al. 1998). In Alston et al. (1998), the private sector contribution in Germany is stated to be 58 percent; however, while their private expenditure data is very similar to ours, they apparently underestimate public agricultural research investments after the reunification.

The breakdown of the R&D ratio by industry reveals the highest research intensities for plant breeding and for the industries where active chemical or biological substances are involved, i.e. crop protection and animal health industry. These branches are becoming more and more knowledge intensive, especially due to the new prospects offered by biotechnology. Some closer insights into the evolution of the plant breeding and crop protection industries, both experiencing significant research-driven restructuring, are given in the following.

Plant Breeding Industry

At the global level, Germany ranks sixth by volume in the agricultural seed market, fourth in terms of seed exports. But unlike other developed countries, where a few large companies often dominate the supply of seeds, in Germany the seed industry is characterized by a large number of small- and medium-sized enterprises. The existing competitive market structure ensures high rates of innovation and a broad biodiversity as measured by the number of available crop varieties. In 1997, there were about one hundred different companies supplying seeds to agricultural producers, fifty-one with their own original breeding programs. The number of companies with their own breeding programs shrank by 23 percent from 1992 to 1997 (see Table 4). However, the research intensity is growing: the total area of plots available for breeding grew by 10 percent and even more impressively, research investments per hectare of breeding plots increased at an annual rate of 2.6 percent. Thus, the R&D ratio in the plant breeding industry rose from 13.3 percent in 1992 to 16.2 percent in 1997 (BDP 1999). Today, the private breeding sector in Germany employs more than 2,000 people, with increasing qualifications, in R&D activities.

One of the main determinants for the rapid structural change in the breeding industry is the advent of biotechnology. Already there are over twenty private laboratories in Germany carrying out R&D with tools of genetic engineering. Due to the substantial investments into biotechnology research, which are necessary to stay competitive in the long run, the individual enterprises are partly abandoning some of their smaller breeding programs (e.g. fodder and oil- and protein-supplying crops) in order to concentrate available resources on their main activities.

Vertical market integration, i.e. large life sciences companies acquiring seed enterprises, became common in many countries but so far affected German plant breeders only to a limited extent. Recently a trend towards horizontal market integration, i.e. mergers between different seed enterprises, is observed. Closer interaction between firms is required in order to get access to the necessary proprietary technological tools and to achieve a minimum critical mass for strategic research.

Table 4:
Structure of the German plant breeding industry (1992 and 1997)

Crop group	Number of companies with own breeding programs		Area of plots available for breeding (ha)	
	1992	1997	1992	1997
Cereals	35	31	1,000	1,060
Maize	18	18	850	1,090
Potatoes	15	13	415	539
Beets	7	6	600	658
Fodder crops	13	4	170	118
Oil and protein suppliers	32	22	415	339
Total[a]	66	51	3,450	3,804

[a] The number of companies does not sum up to the total because most of the companies have breeding programs in more than one crop group.
Source: BDP (1999).

Nevertheless, to date most partnerships between German seed enterprises are through cooperation rather than complete integration. Joint applied research activities involving different private plant breeders as well as public institutes are fairly widespread. Such projects are partly financed through public funds administered by the Society for the Support of Private Plant Breeding (GFP). Moreover, the German government launched a new program on Plant Genome Analysis (GABI) in 1998 aimed at stimulating public–private partnerships in the area of basic research. Additionally, there are numerous private initiatives for international cooperation between breeding enterprises. Ten out of the fifty-one breeding companies in Germany are subsidiaries of foreign firms. Thirty-two enterprises are maintaining extensive cooperation in R&D, seed production and dissemination with foreign partners inside and outside Europe. It appears that German plant breeders are quite successfully managing the challenges imposed by the changing framework conditions. In this respect, the public sector should feel encouraged to learn from the flexibility that the private sector is demonstrating.

Crop Protection Industry

The global crop protection industry is dominated by a comparatively small number of transnational life sciences companies. Three German firms, AgrEvo, Bayer and BASF, rank among the top ten worldwide in terms of aggregate sales. As is the case for plant breeding, the industry became much more knowledge intensive with the incorporation of biotechnology. Therefore the research budgets of crop protection firms are rising continuously. Apart from the European countries, the major export markets of German companies are in North and South America and in Asia. Today, a global outreach requires more than just the positioning of end products in foreign markets. Due to the rising complexity of agronomic problems, site-specific research is gaining in importance. Therefore, German firms maintain an increasing number of research and experiment stations in relevant regions all over the world. Additionally, successful penetration of new markets is facilitated by international alliances and institutional networks. Table 5 shows selected collaborative projects involving German crop protection companies from 1991 to 1998.

Numerous national and international joint ventures have been formed during the last decade. However, Table 5 also shows that acquisitions and mergers have characterized the crop protection industry. In the short to medium run, company mergers can increase shareholder value by exploiting synergistic effects. Yet there are also more long-term aspects involved, mainly spurred by R&D objectives:

- As for plant breeding, international competition requires that more and more basic and strategic research be carried out in the crop protection industry. Basic research is associated with extended time lags between project start and final technology release. The R&D resources of a single company might be too limited to ensure a sufficiently high rate of innovation in the future.
- The international strengthening of IPRs brings about a situation in which more and more essential technology components and

research tools are patented. Mergers between different companies broaden the portfolio of patents available for in-house research.

- For most of the crop biotechnologies, a high-yielding and adapted crop variety is the only means of realizing innovation on a large scale. Since crop protection companies are extending their biotechnology activities, they need access to seed markets for product delivery. This leads to vertical market cooperation and integration.

Table 5:

Selected acquisitions and alliances of German companies in the crop protection industry (1991-1998)

Company	Year	Corporation involved
AgrEvo	1991	Schering acquired Heyboer (The Netherlands)
	1994	AgrEvo formed (60 percent Hoechst, 40 percent Schering); acquired Misung Agrochemicals (South Korea)
	1995	Acquired Kemira Agro's sugar beet herbicide business; joint venture with Tianjin Pesticide Co and Bohai Chemical Industries (China); joins Rhône-Poulenc in Kyrgyz Agribusiness Company
	1996	Acquired Plant Genetic Systems (Belgium); joint venture with Marubeni (Japan); joint venture with Jiangsu Agrochemical Co (China); collaboration with Pioneer Hi-Bred (USA)
	1997	Acquired Sun Seeds (USA); joint venture with Cotton Seed International (USA)
	1998	Acquired Cargill North America; alliance with Gene Logic; announced merger between Hoechst and Rhône-Poulenc to form Aventis
BASF	1995	Joint venture with Nippon Soda and Mitsui & Co (Japan)
	1998	Joint venture with two German public genetic institutes
Bayer	1991	Acquired remaining stake in Agro-Kemi (Denmark)
	1994	Announced plans for a joint venture with Monsanto

Sources: Information provided by Hoechst Schering AgrEvo GmbH, and James (1998).

It remains to be seen how the international markets for seeds and agrochemicals will look in the twenty-first century. While increasing integration is apparently stimulating innovation at present, it must not be overlooked that too high a market concentration might have the opposite effect in the long run. In any case, the broadening of research mandates of the private sector cannot leave public sector research policies unaffected. New options of intersectoral partnerships must be identified.

Contribution to International Agricultural Research

The Consultative Group on International Agricultural Research (CGIAR) with its sixteen centers in different countries is the largest international entity of the global research network dedicated to the agricultural problems of the developing world. The CGIAR was established in the early 1970s, and today it has an overall annual budget of about $330 million US. Part of these funds are raised through international organizations and foundations, yet about two-thirds of the overall budget is sponsored through direct contributions of the various member countries. Europe as a whole, the individual countries as well as the European Commission, accounts for almost 40 percent. With a 6 percent share of the total CGIAR budget, Germany is the single largest European donor country, and at the global level, ranks third after Japan (13 percent) and the USA (12 percent) (BMZ 1998). The German financial support to the CGIAR is administered by the Federal Ministry for Economic Cooperation and Development (BMZ) and the German Agency for Technical Cooperation (GTZ).

However, considering only these regular financial flows neglects other important linkages that exist between Germany and the international agricultural research network. First, additional money is made available by BMZ for short and medium-term agricultural research undertakings that can be implemented by the CGIAR centers but also by other international organizations. Second, there is a substantial exchange of knowledge through German researchers working in the CGIAR centers. Third, many agriculture-related development projects, particularly those implemented by GTZ, have collaborative components with CGIAR centers. Fourth, and probably most important, there are numerous research projects carried out by organizations in the German NARS with a clear focus on aspects relevant to the tropics and subtropics (cf. ATSAF 1997). Many of these research undertakings have close partnerships with CGIAR centers and other international and national organizations. Often this transboundary cooperation works better than the coordination of such efforts among different German research organizations. Of course, the private sector in Germany also contributes substantially to international agricultural research; such technology spillovers were discussed in the previous section.

Conclusions

Agricultural research in industrialized countries is confronted with tremendous changes in national and international framework conditions. On the one hand, the relative importance of the agricultural sector, defined in a narrow sense, is declining, which makes it difficult to justify immutable public research expenditures. But on the other hand, important new and complex topics, such as sustainability issues and altered consumer preferences, emerge which have to be covered by systemic research approaches. Although some argue that agriculture-specific subjects will gradually be absorbed by their parent disciplines, it must not be overlooked that agricultural scientists have an interdisciplinary educational background, making them well-qualified to meet the rising complexity of today's research tasks. New challenges for research also emerge because of globalization tendencies. Markets for commodities and factors of production, including knowledge, are becoming more closely interlinked at the international level. Likewise, the fundamental problems facing humankind, such as hunger, poverty and resource depletion, more and more have a global dimension. Low-income countries, where these problems are most severe, often do not have the scientific capacity to develop needed technologies. So the industrialized countries have the responsibility to fill in the gap. This not only applies to the rich countries' financial and human capital contributions to the international system but also to development-oriented research carried out in their own NARSs in close collaboration with partners from the less developed countries.

The question of effectiveness and efficiency of the German NARS has been raised in the past, but there are no comprehensive economic evaluations of the German agricultural research system available. Also sufficient studies on the rates of returns of major components of the system are lacking. Thus, there are also some doubts about the ability of the present German agricultural research system to efficiently handle the new European and global situation. Changing framework conditions require processes of institutional adjustment. While Germany served as an innovative model for the establishment of agricultural research systems in other countries some one-hundred-fifty years ago, the institutional adaptation processes today are rather slow. Research coordination and evaluation in Germany has to be improved in order to remain competitive in the global research system.

For university research this means that the degree of specialization of each faculty should be increased (cf. Tangermann 1999). With decreasing student numbers, not every faculty needs to offer advanced training and research in all areas of agricultural sciences. It is rather desirable that in the future centers of excellence be established with a better defined focus, to take advantage of positive synergies and scale effects. Greater specialization and coordination would not only avoid inefficient research overlaps but could also increase the faculties' scientific reputation and their research impact. Furthermore, the structure of German university education, including the academic degrees offered, should be adapted to the Anglo-Saxon system to facilitate the international exchange of students. The discussion about closing down individual agricultural faculties has been raised repeatedly in recent times (cf. Block 1996).

The structure of nonuniversity research also needs to be reconsidered. Of course, the claim for greater specialization and the avoidance of inefficient research overlaps applies to university and nonuniversity organizations alike. However, of the federal and state research centers it must also be asked whether all their tasks must be carried out as departmental research. It is hypothesized that the efficiency of the German NARS could be enhanced if only typical sovereign duties were directly assigned to these centers. The allocation of additional funds should be made subject to competition, whereby the departmental research centers should undergo the same independent review processes as other public institutes (cf. Koester and von Cramon-Taubadel 1994). On the other hand, this would also increase the possibility of departmental research centers raising third party funding from sources other than their parent ministries.

In a system restructuring, the changing role of the private sector in agricultural research also needs to be considered. Generally, it is argued that a sufficient amount of public agricultural research investments is required to develop those intermediate and end technologies that would not be provided by the private sector because of market failures. This is true, of course, but it is also important to note that the execution of public research is not the only possible answer to market failures. The establishment of stricter IPR regimes can create new incentives for private research. In the area of biotechnology, for instance, private enterprises increasingly also carry out basic research. So the division of tasks between public and private research becomes blurred. A redefinition of the traditional mandates is required. By nature, private companies respond more flexibly to changing environments. Academic research organizations should also feel encouraged to adopt some private sector research management principles. Most important, however, is that the cooperation between public and private sector research is being strengthened. Such partnerships are vital to get mutual access to proprietary technology components and to efficiently harness the comparative advantages of each sector. The identification of viable IPR systems for public–private research joint ventures is an important undertaking for the future.

It is clear that the implementation of the reform requirements will be a protracted process with numerous obstacles of path dependence and organizational sluggishness. Ironically, the process might be accelerated by the government's budget-saving objectives leading to a reduction of institutional budgets and a greater need to attract third party funding. Yet the German NARS should not only react passively to such exogenous events. A closer cooperation and innovative networking among the different research agents has to be actively sought, which also holds for the various providers of external funds. In the longer run, mergers of different research organizations might be unavoidable to sustain a critical mass for increasingly complex and multidimensional research tasks. Such system adjustments should always be understood positively as an opportunity to increase the domestic and international impact of German agricultural research. Experience with NARS reforms in other countries might help to guide developments in Germany. Alston et al. (1999) present an overview of the ongoing reform processes in different industrialized countries. System restructuring, however, should not only be seen in the national context. The increasing globalization of agricultural and food problems has to be tackled through joint international research efforts.

There is a need for better coordination of agricultural research policies in the EU and at the global level.

References

Abelson, P.H., and P.J. Hines (1999). The Plant Revolution. In: Science, Vol. 285, 16 July, pp. 367–68.

Alston, J.M., P.G. Pardey, and V.H. Smith (eds.) (1999). Paying for Agricultural Productivity. Johns Hopkins University Press, Baltimore.

Alston, J.M., P.G. Pardey, and J. Roseboom (1998). Financing Agricultural Research: International Investment Patterns and Policy Perspectives. In: World Development, Vol. 26, No. 6, pp. 1057–71.

Anderson, J.R., P.G. Pardey, and J. Roseboom (1994). Sustaining Growth in Agriculture; A Quantitative Review of Agricultural Research Investment Patterns. In: Agricultural Economics, Vol. 10, No. 2, pp. 107–23.

ATSAF (1997). Agriculture Between Environmental Concerns and Growing Food Needs. Council for Tropical and Subtropical Agricultural Research (ATSAF), Bonn.

BDP (1999). Geschäftsbericht 1998. Federal Association of German Plant Breeders (BDP), Bonn.

Block, H.-J. (1996). Die Zukunft der Agrarwissenschaftlichen Fakultäten in Deutschland. In: K. Hagedorn (ed.). Institutioneller Wandel und Politische Ökonomie von Landwirtschaft und Agrarpolitik. Campus Verlag, Frankfurt, pp. 309–56.

BMBF (1998). Bundesbericht Forschung; Faktenbericht 1998. Federal Ministry for Education and Research (BMBF), Bonn.

BML (1999). Forschung im Geschäftsbereich des Bundesministeriums für Ernährung, Landwirtschaft und Forsten. Federal Ministry of Agriculture (BML), Bonn.

BML (1998). Forschungsrahmenplan. Federal Ministry of Agriculture (BML), Bonn.

BML (various issues). Agrarbericht der Bundesregierung. Federal Ministry of Agriculture (BML), Bonn.

BMZ (1998). International Agricultural Research for a Secure Future; The German Contribution to the CGIAR System. Federal Ministry for Economic Cooperation and Development (BMZ), Bonn.

von Braun, J. (1999). The Present Situation of Agricultural Research in Germany. In: Agrarspectrum, Vol. 29, DLG-Verlag, Frankfurt, pp. 8–16.

von Braun, J. (1998). The New International Economic Context for European Agricultural Research. In: G. Paillotin (ed.). European Agricultural Research in the 21st Century. Springer, Berlin, pp. 5-12.

von Braun, J., and P. Wehrheim (1995). Agrarforschung und Agrarentwicklung: Wirkungen und Politikorientierung. In: German Association of Agricultural Economists (GEWISOLA). Proceedings of the Annual Conferences, Vol. 31, Landwirtschaftsverlag, Münster, pp. 565-79.

Brinkmeyer, D. (1996). Transaktionskosten der öffentlichen Projektförderung in der deutschen Agrarforschung. Shaker Verlag, Aachen.

Burian, E. (1992). Staatliche Agrarforschung: Eine vergleichende Analyse für Frankreich, das Vereinigte Königreich und die Bundesrepublik Deutschland. Agrarwirtschaft, Sonderheft 136, Verlag Alfred Strothe, Pinneberg.

Byerlee, D. (1998). The Search for a New Paradigm for the Development of National Agricultural Research Systems. In: World Development, Vol. 26, No. 6, pp. 1049-55.

Csaki, C. (1998). Agricultural Research in Transforming Central and Eastern Europe. In: European Review of Agricultural Economics, Vol. 25, No. 3, pp. 289-306.

FEDESA (1997). Animal Health; Facts and Figures about the European Animal Health Industry. Dossier 14, European Federation of Animal Health (FEDESA), Brussels.

GEWISOLA (various issues). Proceedings of the Annual Conferences. German Association of Agricultural Economists (GEWISOLA), Landwirtschaftsverlag, Münster.

Henze, A. (1997). Ansätze zur Effizienzsteigerung im Hochschulbereich; Generelle Reformansätze und Anpassungserfordernisse für die Agrarwissenschaften. In: Agrarwirtschaft, Vol. 46, No. 11, pp. 384-98.

Hockmann, H., and G. Recke (1997). Public Agricultural Research in Germany. Paper Presented at the 51st EAAE Seminar "Innovation for Innovation; The Organization of Innovation Processes in Agriculture", 21-23 April 1997, Zandvoort.

Isermeyer, F., and W. Werner (1997). Perspektiven der deutschen Agrarforschung; Zusammenfassende Überlegungen. In: Agrarspectrum, Vol. 26, DLG-Verlag, Frankfurt, pp. 161-80.

IVA (1998). Jahresbericht 1997/98. Association of the Agricultural Industry (IVA), Frankfurt.

James, C. (1998). Global Review of Commercialized Transgenic Crops: 1998. ISAAA Briefs No. 8, International Service for the Acquisition of Agri-biotech Applications, Ithaca, NY.

Koester, U., and S. von Cramon-Taubadel (1994). Zur Struktur und Organisation der Agrarforschung in der BRD. In: Agrarwirtschaft, Vol. 43, No. 8/9, pp. 293-95.

LAV (1998). Jahresbericht 1998. Association of the Agricultural Machinery Industry (LAV), Frankfurt.

Lückemeyer, M. (1999). Prospects for Efficient Agricultural Research in Germany as Seen by Non-University Research. In: Agrarspectrum, Vol. 29, DLG-Verlag, Frankfurt, pp. 17-24.

Müller, R.A.E., and D. Brinkmeyer (1996). Organisation von Forschungsprojekten in der Agrarforschung. In: Agrarspectrum, Vol. 25, DLG-Verlag, Frankfurt, pp. 164-80.

Paillotin, G. (ed.) (1998). European Agricultural Research in the 21st Century. Springer, Berlin.

Reichrath, S. (1990). Entstehung, Entwicklung und Stand der Agrarwissenschaften in Deutschland und Frankreich. Ph.D. Dissertation, University of Kiel.

Reisch, E. (1999). Entwicklungslinien der agrarökonomischen Forschung in der Bundesrepublik Deutschland, 1959-1999. Paper presented at the 40th Conference of the German Association of Agricultural Economists (GEWISOLA), Kiel.

Rosenkranz, O. (1999). Entwicklungslinien der agrarökonomischen Forschung in der Deutschen Demokratischen Republik, 1959-1989. Paper presented at the 40th Conference of the German Association of Agricultural Economists (GEWISOLA), Kiel.

Ruttan, V.W. (1982). Agricultural Research Policy. University of Minnesota Press, Minneapolis.

Tangermann, S. (1999). Perspectives for Efficient Agricultural Research in Germany as Seen by the German Science Council. In: Agrarspectrum, Vol. 29, DLG-Verlag, Frankfurt, pp. 25-34.

Virchow, D. (1999). Conservation of Genetic Resources; Costs and Implications for a Sustainable Utilization of Plant Genetic Resources for Food and Agriculture. Springer, Berlin.

Chapter 13
The Value-added Chain in the German Food Sector

Hannes Weindlmaier
Dairy and Food Research Centre Weihenstephan,
Technical University of Munich

Introduction

Given the current conditions of agriculture, it must be recognised that the success of the sector does not chiefly depend on efficient production and adequate structure of agricultural enterprises. In particular the concept of Total Quality Management is focussing on the fact that the agro-food chain will only be successful if each component works well. On the one hand, the degree of satisfaction the consumer derives from products he buys depends not primarily on agriculture but on decisions made by the downstream elements of the food chain. The product characteristics consumers expect must be traced from distribution and processing back to the producer of the raw materials. On the other hand, for farmers it is vitally important that the downstream subsectors of the food chain are able to transform agricultural products into value-added products which can be sold at a reasonable price.

The enterprises in each subsector along the food chain have their specific functions and problems. The strategies of the actors in the production area and distribution channel are partly different and sometimes contradictory. There are also significant differences in the structure of enterprises in the different subsectors of the food chain. Despite insufficient structures, farmers' bargaining power has developed quite satisfactorily since World War II. Recently however, their marketing power has started to gradually decline due to the increasing degree of concentration in the food processing industry, and especially in the food retail trade. The surplus situation for most food products and the policy decisions made related to the CAP and WTO are further explanations for farmers becoming the weakest element in the food chain.

In this chapter the characteristics and developments of the different subsectors of the value-added chain in the agricultural and food sector will be analysed and major problem areas will be addressed. In this context the emphasis will be put on the most important subsectors besides agriculture, namely the food industry and the food retail trade (for an analysis of the agricultural sector see chapter 2 of this book).

Structure and Subsectors of the Value-added Chain in the Agricultural and Food Sector

The Structure of the Value-added Chain

The value-added chain in the agricultural and food sector can be divided into several subsectors (see also Besch, 1993). The left side of figure 1 shows agricultural production at the beginning of the food chain. In 1997 sales for the agricultural sector amounted to 60.8 billion DM in Germany. For further processing, the agricultural goods then are assembled by the wholesale trade, by the food processing industry, and/or by food handicraft enterprises. Only a small percentage of agricultural products is directly sold to private and bulk consumers. While the food handicraft enterprises sell most of their products directly to consumers, the industrially processed food reaches the final consumer via the food wholesale and/or retail trade. The total value of private consumption of food, drinks, and tobacco totals 377 billion DM in 1997 compared with 340 billion DM in 1991.

The Value-added Chain in the German Food Sector

Figure 1:
Structure and components of the value-added chain in German agriculture[1]

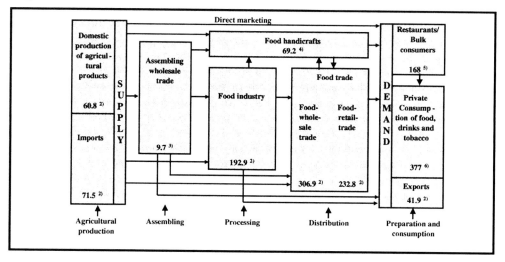

[1] In general, the numbers (billion DM, without value-added tax) refer to 1997. The numbers of the assembling wholesale trade refer to 1997/98.
[2] Source: Statistisches Bundesamt (1999).
[3] Source: Ernährungsdienst (1999).
[4] Sources: Statistisches Bundesamt (1999); Deutscher Fleischer-Verband (1988); Zentralverband des Deutschen Bäckerhandwerks (1998).
[5] Source: CMA Mafo Briefe (1999).
[6] Private consumption of food, drinks and tobacco in billion DM. Source: Statistisches Bundesamt (1998: 664).

Table 1 shows that the German food sector has been quite successful during the last decades in providing food for the continually growing German population; in spite of 4.4% population growth during this period, for most product groups the degree of self-sufficiency increased quite remarkably, especially for cereals, potatoes, sugar, and milk products. On the right the table shows marketing margins calculated as the difference between farm sales and consumer expenditures for food of domestic origin (see also Besch, 1993: 29-31). It is obvious that the farmer's share in consumer food expenditure is constantly decreasing while marketing margins have increased accordingly. This development can be observed for all food products with varying intensity, although the decrease of the farmer's share becomes especially evident with regard to vegetal products and meat.

Table 1:
Indicators for the development of the German food sector, 1975/76 to 1996/97

Product / group of products	Degree of self sufficiency (%)			Share of agricultural sales in % of consumer expenditures for products of domestic origin		
	1975/76	1992/93	1996/97	1975/76	1992/93	1996/97
Cereals	81	117	120	18.8 [2]	6.8 [2]	4.5 [2]
Potatoes	90	97	102	53.4	23.0	19.2
Sugar Beets and sugar	110	141	150	42.3	37.1	38.2
Meat and meat products [1]	82	82	85	51.7	28.7	30.4
Fresh milk and fresh milk products	100	108	116	60.7 [3]	48.7 [3]	45.7 [3]
Cheese	91	96	102			
Butter	136	87	77			
Eggs and products made of eggs	80	75	72	85.2	68.0	74.7

[1] Data for animal products refer to calendar years, that means 1975/76 refers to 1976.
[2] Data refer to cereals for bread and products made of bread grains.
[3] Data refer to milk and milk products.
Source: Statistisches Bundesamt and Bundesministerium für Ernährung, Landwirtschaft und Forsten.

According to Wöhlken (1991: 123), the increase of marketing margins is due both to the range of built-in services and the costs of these services. The changes in the marketing margins clearly indicate the shift of the added value within the food chain in recent decades.

Importance and Development of Agricultural and Food Imports and Exports for the German Food Chain

Imports and exports of commodities and processed food play an important role in Germany as can be seen from figure 1. Imports of commodities and processed food amount to 71.5 billion DM in 1997. Worldwide Germany ranks first place importing agricultural and food products. Major imported products are fruits and vegetables, meat and fish, cacao products, dairy products, coffee, and tea. However, Germany also ranks fourth worldwide in exporting agricultural and food products. Between 1991 and 1997 exports increased from 35.8 billion DM to 41.9 billion DM. During this period the negative trade balance decreased from 32.1 to 29.6

billion DM. Important export goods are dairy products, fresh and processed meat, fruits and vegetables, tobacco, and tobacco products.

German trade in agricultural and food products is primarily concentrated on the intra-EU trade: in 1997 62.6% of the imports and 67.9% of the exports were realised with EU Member States. During the 1990s, exports to East Europe increased remarkably. Although this positive development has been interrupted due to the economic crisis in these countries, it is expected that in the middle and long run these countries will be important partners for the German food sector.

Importance and Characteristics of Cooperatives and the Assembling Wholesale Trade

The assembling wholesale trade is an important intermediary, especially for vegetal products (42% of sales in 1997/98), while animal products are often assembled by the food industry itself. The assembling wholesale trade consists of two legal forms. In 1997/98 about 1,070 private wholesale companies and 1,205 cooperatives were buying products from farmers, but also supplying them with feedstuffs, fertiliser, pesticides/insecticides, and agricultural machinery (Ernährungsdienst, 1999). In general, deliveries to farmers account for about 70% of total sales of the assembling wholesale trade.

During recent decades, the number of companies in the assembling wholesale trade significantly decreased, affecting both the private and the cooperative sectors. The decreasing number of farms coupled with economic pressure were the main driving forces for this concentration process.

Cooperatives play a very important role in addition to their wholesale trade in providing inputs to farmers and assembling agricultural products. Traditionally, cooperatives are also active in processing agricultural products, specifically in the dairy sector, the slaughterhouse sector, in the processing of fruits and vegetables, in wineries, and in the production of feed mixes. While in 1960 20,926 cooperatives with 3.6 million members served the agricultural and food sector in Western Germany (Deutscher Raiffeisenverband, 1999), there were less than 4,221 cooperatives with about 3 million members in the reunified Germany in 1998. The total turnover of cooperatives involved in assembling and processing activities amounted to 75 billion DM in 1998.

Recently, cooperatives in some areas have lost market share to capital-oriented forms of enterprises such as Plc (AG) and Ltd (GmbH). Cooperatives are facing several problems. One of their major shortcomings is connected with the procurement of equity capital. Because of the actual income situation, the farmer-members are often economically unable or unwilling to increase their shares or to provide additional needed equity capital to finance growth and long-term strategies (Weindlmaier, 1994). Another problem is associated with the structure of the managing and supervisory boards of the cooperatives. By law they may consist only of members of the cooperative. Therefore, quite often decisions are strongly influenced by the members' short-term objectives, i.e. to gain high prices for the products they supply instead of being oriented toward long-term strategic needs of the processing cooperative. Hence, deficiencies with respect to investments in marketing activities like branding, product innovation, advertising, and internationalisation are frequently found in German cooperatives.

The cooperatives' union is aware of these problems. In a joint initiative with the German farmers' union, new strategies and measures were formulated and discussed. Their objective is to regain the position the cooperative sector formerly had in the German food chain (Deutscher Raiffeisenverband e.V. et al., 1998).

Contract Farming and Vertical Integration

Some subsectors of the food industry are confronted with high market risk resulting from inelastic supply and high fixed production costs. In these fields contract farming has a long tradition in Germany (Besch, 1993: 25). Contracts should guarantee that the products delivered by the farmers meet the processors' needs with regard to quantity, quality, time of delivery, etc. Contract farming plays an important role in the vegetable, potato processing, and sugar industry as well as in the canning industry. Furthermore, contract farming is common between the processing industry and poultry and milk producers.

Vertical integration also became an important strategy in integrated production and marketing programs, especially for branded meat during the last years. These programs, based on modern quality management standards (such as ISO 9000), aim at improving and guaranteeing the quality of the final products by preventive controls through the entire production and marketing process (Weindlmaier et al., 1997). The BSE crisis and quality problems with pork and veal have marked the introduction of these programs. Ottowitz (1997) reported 94 branding programs for beef and/or pork in 1996.

Despite these developments some experts criticise that in other European countries (e.g. The Netherlands, France, Denmark and the United Kingdom) contract farming and vertical integration is much more widespread than in Germany (Kallfass, 1993). Consequently there are competitive disadvantages for the German food sector.

The establishment of producer groups according to the market structure law of 1969, mark another institutional development at the interface between agricultural production and food processing. Their formation should help to adjust the disequilibrium between farmers' fragmented supply of goods on the one hand and increasing concentration in the subsequent sectors on the other. The producer groups' objective is to improve the farmers' bargaining power and to collect and standardise the small supply quantities of individual farmers. To encourage the establishment of producer groups, subsidies are provided by the government.

The number of governmentally recognised producer groups in Western Germany peaked in 1990 with 1,479 groups. Since then, the number decreased to 1,027 in 1997. However, in the 1990s an additional 202 producer groups were founded in Eastern Germany. The market shares of producer groups in Western Germany in 1996 totalled 39.8% for wine, 21.9% for potatoes, 17.6% for slaughter pigs, 16.5% for cereals, and 10.3% for slaughter cattle. In Eastern Germany relatively high market shares have been reached for slaughter animals, i.e. 48.5% for pigs, 38.3% for cattle (Bundesregierung, 1998: 12-13).

According to the results of analyses by Elsinger (1991) producer groups have not yet been able to satisfy the high expectations of their promoters with respect to improving the farmers' marketing position. Instead, producer groups are quite often additional competitors, only increasing regional price struggles. The main reasons for this development are the insufficient market shares of producer groups, the fact that members often do not offer their total production to the producer group but sell to other buyers as well, and management deficiencies. In spite of these shortcomings, experts still assign producer groups an important contribution in the necessary process of intensifying vertical coordination in the German food chain (Halk, et al. 1999).

Direct Marketing of Agricultural Products

While most farm output is processed and reaches the consumers via the trade channel, for some products (potatoes, eggs, apples, milk, wine, and vegetables) direct marketing to consumers has gained some importance. From the farmers' point of view the potential advantages of direct marketing are higher product prices, lower price variability, and the possibility of employing excess labour capacity at the farm. Prerequisites of direct marketing are a farm location which can be easily reached by a sufficient number of consumers, excess labour capacity, entrepreneurial skills, and marketing measures specifically designed for the target groups.

Table 2:

The development of direct marketing shares of potatoes and eggs, 1991/92 to 1996/97

Product	1991/92	1992/93	1993/94	1994/95	1995/96	1996/97
Potatoes	21.5	26.7	26.4	27.2	17.9	19.1
Eggs to bulk consumers	28.9	22.4	22.0	20.4	18.0	17.8
Eggs to individual consumers	19.4	8.0	7.2	6.9	6.7	6.7

Source: Statistisches Bundesamt (1999: 174).

Wirthgen and Maurer (1992) estimated that at the beginning of the 1990s only about 5% of German farmers were active in direct marketing. In a recent study Wirthgen and Kuhnert show that the share of direct marketing sales in total sales of farms accounts for about 9% (Kuhnert, 1998: 59-76). According to this study about 70% of the farmers who are active in direct marketing offer processed food as well. Table 2 shows the development of the market shares for the two main directly marketed goods, i.e. potatoes and eggs. The figures point out that the percentages vary year by year depending mainly on the annual yields. The numbers indicate however, that direct marketing of these product groups is gradually decreasing.

Currently a renaissance of direct marketing is seen with respect to the sale of biological products. It is estimated that about 80% of farmers producing according to specific ecological criteria are involved in direct marketing. The products offered include grains, potatoes, fresh vegetables, fruits, milk, etc.

Role of Food Handicraft Enterprises in the Value-added Chain of German Agriculture

Food handicraft enterprises still have a relatively high importance in Germany. They offer a wide range of high quality products and regional specialities. As there is no strict legal differentiation between the food industry and food handicraft enterprises, it is statistically impossible to exactly split these two sectors.

In 1997 butchers had a turnover of 35.9 billion DM. Their market share of the total sales of meat and meat products is about 50% (Deutscher Fleischer-Verband, 1998). In spite of the rapid ongoing concentration process leading to a reduction in the number of enterprises, the competitiveness of butchers improved in recent years. This is primarily caused by the BSE crisis and by a diversification of their activities. Many butchers established branch outlets. In addition to selling meat and meat products to consumers, butchers increasingly use additional market channels. In 1997 84% of the butchers offered a party/catering service. Furthermore, more than half of the butchers are also selling products to bulk consumers like restaurants and nearly half of them offer fast food.

With regard to market share, bakeries have been even more successful although the concentration process continues rapidly. Turnover for bakeries amounted to 26.5 billion DM in 1997. The market share of bakeries in the German bread market is still about 65 to 70% (Zentralverband des Deutschen Bäckerhandwerks, 1998). Based on numerous innovations, bakeries are offering a diversified range of bakery products.

Developments in the German Food Industry

Changes in the Underlying Conditions for the Food Industry

The conditions for the food industry significantly changed in recent years. In figure 2 some of the most important conditions for the food industry are listed. A first important factor is the growing internationalisation of markets. This process has accelerated by the Single European Market established in 1993 which aims at full integration of goods, capital, services, and labour. For the food industry this has led to a further increase of competition within the EU, and especially in Germany. The large potential market of about 80 million consumers, the high purchasing power of German consumers, and the good infrastructure are the main reasons for this effect. Even companies which mainly sell their products regionally or nationally are confronted with a growing number of competitors. Therefore, further internationalisation is also a very important strategic objective for German food enterprises. Another important development is the harmonisation of legal conditions for the food industry (e.g. the food law, the transportation law, the trademark law, etc.) in the EU (see Wendt, 1991). In recent years the food industry has been forced to make many adjustments to be in line with the new legal framework. Also of great importance is the fast-growing concentration process in the food trade and the changes in consumer behaviour and food demand (see Nienhaus, 1995; Nestlé, 1999).

Recent trends in consumer attitudes and behaviour are the growing polarisation of consumer requirements (the so called hybrid consumer), a constantly high consideration of the quality, taste, and freshness of food, and consumers highly sensitive to raw materials and additives used, techniques employed for processing, and food safety. Furthermore consumers show growing sensitiveness with respect to the price-quality relationship. Even for people with high incomes, price is an important buying criteria; they seek to buy high quality brands as cheaply as possible ('smart shoppers'). It can also be noted that the regionalism regained importance for some consumers.

Figure 2
Conditions for the German food industry

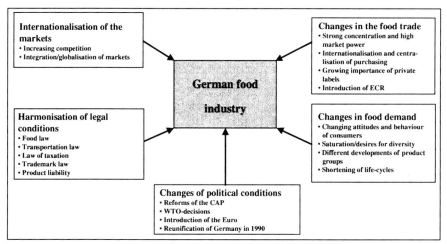

These behavioural changes led to modifications in food demand and to new challenges for the actors in the food chain. The general saturation of food demand on the one hand and the consumers' longing for variety and novelty on the other is remarkable. The demand for basic food is decreasing while the demand for value-added products (convenience products, for biological, ethnical, or functional food) is steadily growing. The life cycle of food products, however, steadily diminishes. Meanwhile out-of-home consumption has become an important issue in food demand; it is estimated to have a market share of more than 40% in Germany (CMA Mafo Briefe, 1999). In general, in the demand-driven food market of Germany, only those food suppliers who strongly focus on consumers' desires and who manage to consider those desires in product development and marketing will be successful.

Last but not least there have been major changes in the general policy framework of the food industry. The changes in the CAP, the Agenda 2000, and the developments in GATT and the WTO are just some important issues (for further details see the policy-oriented chapters of this book). It is expected that the introduction of the euro will lead to further price competition in the European food sector.

A change in the policy framework of specific importance for the German food industry has been the reunification of Germany in 1990. In 1991 the food industry in Eastern Germany had an annual turnover of about 16.7

billion DM which means that about 7.8% of the total turnover of the German food industry were realised by Eastern German companies (Bundesministerium für Ernährung, Landwirtschaft und Forsten, 1993: 116). The main part of the East German food industry was not competitive under Western economic conditions. The main reasons for this were marketing deficits, obsolete buildings and equipment only enabling very low productivity, severe shortcomings in the quality of products, and a product assortment which did not meet modern consumer demands.

Since reunification, an enormous selection process has taken place in the food industry of Eastern Germany. Many companies were confronted with financial problems and had to be liquidated. Modern processing facilities have been built. Often, the cooperation with partners from Western Germany or other Western countries resulted in these partners acquiring the capital majority or becoming the principal owner of the enterprises. This restructuring program has been supported with subsidies by the Governments of the Laender, by the federal government, and by the European Union totalling up to 50% or more of total investment costs. Nowadays the food industry of Eastern Germany is one of the most modern in Europe.

Important Sectors of the Food Industry and their Structure

Table 3 shows the most important branches of the German food industry ranked by their turnover. By far the most important branch is the dairy industry with an annual turnover of about 40 billion DM. It is followed by the industrial preparation and preservation of meat and meat products, the manufacturing of malt and beer, and of bread, pastry, and cakes. In 1998, about 16.4% of the turnover was realised with exports. The number of enterprises in the food industry is steadily decreasing because of the extensive concentration process. In 1998 altogether 5,911 enterprises were counted.[1] If we look at figure 3 we notice that the concentration levels in

---

[1] Due to changes in the statistical sample in 1997 it is not possible to show the long-term development.

the different branches vary considerably. In this figure, two different methods to measure the concentration rate were employed.

Table 3:
Turnover and employment in the German food and drink industry, 1998

Rank	Area of economic activity	Turnover (billion DM)	(%)	Employees
1.	Dairies, cheese making, preserved milk products and processed cheese	39.626	17.38	43,901
2.	Industrial preparation and preserving of meat and meat products	23.353	10.24	84,153
3.	Manufacture of malt and beer	19.501	8.55	42,689
4.	Manufacture of bread, pastry and cakes	16.927	7.42	150,481
5.	Slaughterhouses	15.749	6.91	23,696
6.	Manufacture of cocoa, chocolate and confectionery	14.263	6.26	32,294
7.	Processing and preserving of fruit and vegetables	13.156	5.77	27,674
8.	Production of mineral water and lemonades	11.573	5.08	23,915
9.	Manufacture of animal feedstuffs	9.808	4.30	11,660
10.	Manufacture of distilled potable alcoholic beverages and wines	9.051	3.97	8,236
11.	Processing of coffee and tea	8.945	3.92	6,239
12.	Manufacture and refining of sugar	6.258	2.75	6,981
13.	Other	39.765	17.44	82,390
	Food and drink industry	227.975	100.00	544,309

Source: Bundesvereinigung der Deutschen Ernährungsindustrie (1999: 13).

Figure 3:
Concentration measures of enterprises in German food manufacturing

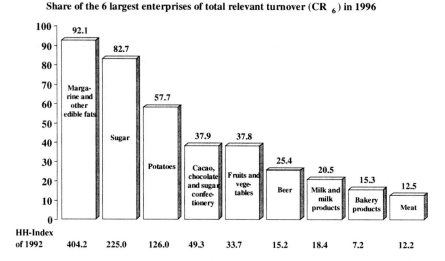

Source: Bundesministerium für Ernährung, Landwirtschaft und Forsten (1999)

First, for some important branches the Concentration Rate 6 (CR$_6$) is shown representing the market share of the six largest enterprises in the branch with respect to total turnover. In addition, at the bottom of figure 3 the Herfindahl-Hirschmann Index for 1992 is presented (Herrmann et al., 1996: 193).[2]

The figure shows that in some branches the largest six companies have already attained an extremely high market share. However, the Herfindahl Hirschmann indices indicate that even in these branches many small companies yet exist. A relatively high concentration rate can be found in branches with a high potential for realising economies of scale, a rather homogeneous product range, and few processing stages. Examples are the manufacturing of margarine and sugar. Concentration is still small in branches with a prevailing regional orientation, which is true for the manufacturing of beer and, at least in some regions, of milk. Concentration is also minor in branches with a long food handicraft tradition.

---

[2] In the Herfindahl-Hirschmann Index, 1,000 indicates maximal concentration.

Table 4:
Top twenty enterprises in the German food industry

No.	Firm / Firm group	Turnover (billion DM)	Main product areas
1	Tschibo Holding AG, Reemtsma, Eduscho	18.600	Coffee, cigarettes, nonfood
2	Südzucker-Gruppe Südzucker AG; Schöller	8.800	Sugar, ice cream, sweeteners, frozen food, biscuits and cakes
3	Oetker-Gruppe; Binding-Gruppe, Henkell & Söhnlein	8.197	Frozen food, beer, wine, ice cream, champagne, liquors
4	Campina Melkunie b.v.	8.100	Dairy products
5	Nestlè-Gruppe Deutschland	7.850	Coffee, baby food, chocolate and confectionery, beverages, meat products and sausages
6	Deutsche Unilever	7.660	Margarine, dairy products, sausages, frozen food
7	Coca-Cola Germany	6.300	Beverages
8	Procter & Gamble Germany	5.900	Detergents, cosmetics, fruit drinks
9	Nordmilch eG	4.700	Dairy products
10	Kraft Jakobs Suchard Germany	4.200	Chocolate and confectionery, coffee, cheese
11	Pfeiffer & Langen-Gruppe, Intersnack; Krüger	3.680	Sugar, snacks, instant products
12	Moksel-Konzern	3.320	Meat and meat products
13	CG Nordfleisch-Konzern	2.800	Meat and meat products
14	Südfleisch-Konzern	2.737	Meat and meat products
15	Humana Milchunion	2.700	Dairy products
16	Schwartau-Gruppe	2.660	Marmalade, beverages, cereal products, baby food, meat
17	Haribo-Gruppe	2.600	Confectionery
18	Ferrero Germany	2.600	Confectionery
19	Melitta-Gruppe	2.240	Coffee, nonfood
20	Molkerei Müller	2.230	Dairy products

Source: Lebensmittelzeitung (1999a: 10

Typical examples are the preparation and preservation of meat by butchers and the manufacturing of bread, pastry goods, and cakes.

Even in some branches that show low concentration rates in figure 3, we recently observe a rather high growth in concentration. This is true, for example, for dairies and cheese production, for the manufacturing of prepared foodstuffs, for slaughterhouses, for the manufacturing and refining of sugar and of macaroni and noodles. This rising concentration may reflect the pressure in these markets to make efforts in realising economies of scale and in increasing market power. Experience shows, however, that despite these concentration processes, the food industry was unable to significantly improve bargaining power.

Table 4 shows the top twenty enterprises in the German food industry in 1998/99. As can be seen, the top enterprises like Tchibo, Südzucker, and Oetker are conglomerates producing a variety of food products. Furthermore, among the largest companies are subsidiaries of multinational food companies like Campina, Nestlé, Unilever, Coca-Cola, Procter & Gamble, and Kraft Jakobs Suchard. The top ten enterprises of the German food industry have a market share in terms of total turnover of food of about 35%. In comparison, the top ten enterprises in the German food trade have a market share of about 90% of total food sales.)

Competitiveness of the German Food Industry

The competitiveness of the German food industry plays an important role for the further development of the value-added chain. Yet, recent investigations regarding competitiveness in Germany draw a picture with lights and shadows (Hartmann, 1993; Weindlmaier, 2000). Table 5 shows the values for the Relative Export Advantage Index (RXA), the Relative Import Penetration Index (RMP), the Relative Trade Advantage Index (RTA), and the ratio of unit-values of exports and imports for selected product groups (Weindlmaier, 2000).[3] Positive ratios of the unit-values indicate

---

[3] Values greater than 1 for the RXA and RTA are interpreted as relative competitiveness compared with other sectors of an economy while values less than 1 are indicators for the opposite. For the RMP, values less than 1 mean high market shares of imports and values

that primarily high value-added products like specialities and branded products are exported while commodities are imported. The calculations are based on the COMTEX database of the EU and refer to intra-EU trade only.

The numbers of the indicators RXA, RMP, and RTA in table 5 emphasise the competitiveness of branches like the processing of liquid milk, yoghurt and other fresh dairy products, skimmed milk powder, sausages, and cocoa mass. For most of the other products the numbers point out deficits in competitiveness. High ratios of unit values can be found for the same products but also for butter, for pigs, pork and processed products of pork, and for cacao products. These results, which are derived from trade statistics, are confirmed by the analysis of the competitive potential and competitive processes according to Porter's diamond. There are several facts that indicate a high competitive potential of the German food industry but there are also significant disadvantages as table 6 shows. Important advantages are the excellent endowment of production factors, the good infrastructure, the large home market with high purchasing power and demanding and sophisticated consumers, the good domestic image of German food, and the existence of competitive suppliers in the related and supporting industries.

Disadvantages for the German food industry are high labour costs and social security charges combined with rigid labour market regulations and small wage differentiation, deficits in the strategic orientation and strategic marketing in the food industry, structural deficits with respect to the size of companies, plants, and farms, and high taxes and public charges.

---

greater than 1 point at low proportions of imports in the relevant product area, i.e. high competitiveness of domestic products.

Table 5
Measures of the competitiveness of the German food industry

Product group	RXA		RMP		RTA		Unit Values	
	Ø 88–90	Ø 95–97	Ø 88–90	Ø 95–97	Ø 88–90	Ø 95–97	Ø 88–90	Ø 95–97
Dairy products								
Liquid milk	1.15	2.25	0.12	0.03	1.03	2.22	0.84	0.76
Cream	2.93	1.06	0.23	0.28	2.70	0.79	0.83	0.95
Skimmed milk powder	3.07	3.51	0.35	0.18	2.73	3.33	0.87	1.03
Yoghurt, buttermilk and kefir	1.89	1.88	0.22	0.27	1.67	1.61	1.53	1.08
Hard cheese	0.92	1.01	0.43	0.61	0.49	0.40	0.81	0.86
Semihard cheese	0.41	0.65	2.97	2.38	−2.55	−1.73	0.97	0.86
Soft cheese	0.45	0.21	2.78	4.54	−2.33	−4.33	0.86	0.93
Quark	1.10	1.01	1.52	1.56	−0.43	−0.56	0.55	0.61
Butter and butterfat	0.22	0.25	0.96	1.30	−0.74	−1.05	1.02	1.04
Pigs, pork and processed products	0.20	0.23	1.24	1.58	−1.04	−1.36	1.21	1.18

Table 5 continued

Product group	RXA		RMP		RTA		Unit Values	
	Ø 88–90	Ø 95–97	Ø 88–90	Ø 95–97	Ø 88–90	Ø 95–97	Ø 88–90	Ø 95–97
Sausages	1.16	1.14	0.85	0.92	0.31	0.21	0.84	0.91
Slaughter cattle, calves, beef, veal, and processed products	0.78	0.74	0.56	0.52	0.22	0.22	0.94	0.89
Processed and canned products of beef	0.02	0.15	1.41	0.63	−1.39	−0.48	0.76	0.89
Vegetable fats and oils for human consumption	0.65	0.49	0.66	0.41	−0.01	0.08	0.76	0.68
Margarine etc.	0.80	0.70	0.30	0.47	0.49	0.23	1.09	0.89
Cacao products (except cacao beans and -butter)	0.68	0.87	0.79	0.86	−0.11	0.01	0.98	1.05
Chocolate and confectionery	0.57	0.82	0.87	0.95	−0.30	−0.13	1.01	1.06
Cacao mass	4.37	2.66	0.26	0.33	4.12	2.33	1.16	1.21

Source: Weindlmaier (2000).

Additionally past agricultural and food policies have hindered the development of appropriate measures and strategies to improve competitiveness.

Table 6:
Competitive advantages and disadvantages of the German food industry

Competitive advantages of the German food industry	Competitive disadvantages of the German food industry
Good education, professionality, and motivation of employees	High costs of labour and social security
Availability of capital at low interest rates	Rigid labour market regulations and small wage differentiation
	High costs of energy
Very good infrastructure	Limited availability of venture capital
Great importance of measures for the protection of the environment	Partial lack of professionality in decision making
Extensive competition, demanding and sophisticated consumers	Deficits in the strategic orientation of the food industry and strategic marketing
Good domestic image of German food	
Large home market and high purchasing power at the domestic market	Structural deficits with respect to the size of companies and plants
Foreign markets with fast growing food demand in surrounding countries	Poor farm structure and comparatively high costs in the production of agricultural raw materials
Production of high quality raw materials by German agriculture	
	High taxes and public charges
Competitive suppliers in the related and supporting industries	Policy measures (CAP, etc.) which hindered the development of appropriate activities and measures to improve competitiveness

Source: Weindlmaier (2000).

The Value-added Chain in the German Food Sector

Food Retailing in Germany

Characteristics and Developments of the German Food Retail Trade

During the last few decades organisational and technical changes took place in the German food retail trade transforming the originally very traditional and small-scale food trade to a modern and highly competitive distribution system.

Rapid and extensive concentration took place in the German food retail sector. Figure 4 shows that the top ten companies have gained a market share of 90%. These concentration processes are connected with the implementation of new internal forms of organisation and changes in the vertical structure of the food trade. Up to the 1970s, food retail trade was organised in a three-stage structure consisting of the central wholesale trade, the regional wholesale trade, and the local retail trade; the cooperatives Edeka and Rewe were typical examples of this structure. With a few exceptions, these levels have been integrated with enterprises combining wholesale and retail trade and centralising purchasing activities. Very high market power of the leading retail trade companies is an important consequence. Aldi and Tengelmann are typical for this kind of food trade enterprise in Germany (see also Lingenfelder and Lauer, 1999).

The total number of retail outlets decreased from 161,359 in 1960 to 74,577 in 1997. In the same period the system of offering food changed. Compared to only 14% in 1960, by 1997 85% of the outlets were self-service stores with a market share in terms of turnover of 99%. The typical supermarket lost market share while large self-service department stores and low-price discount stores quickly gained importance. In 1997 about 25% of sales were realised by large self-service department stores and 31% by low-price discount outlets. This development is specifically important in that for these types of outlets, low prices are an important selling point. The increasing introduction of private labels supports the low price strategy. Recently this has become even more important because of intense price struggles between some of the low-price chains like Aldi and Lidl and the US trade giant Wal-Mart, engaged in Germany since 1997. Consequently, the whole food trade had to react to the strategies of these competitors who attempt to realise their low price strategy primarily by low purchasing prices. Extraordinary price pressure on the food industry and other suppliers was the consequence.

Figure 4:
The top 20 enterprises in German food retail trade in 1998

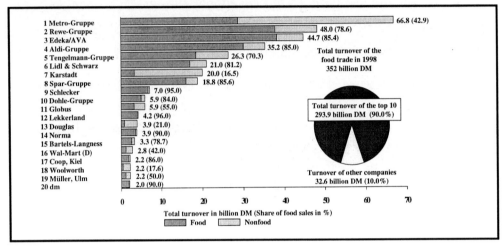

Source: Lebensmittelzeitung (1999b: 4).

In addition to the concentration, strong internationalisation processes are seen in the German food retail trade. For example, the largest German food retail enterprise, Metro, has affiliates in seventeen European countries. The leading German food trade enterprises (e.g. Metro, Rewe, Aldi, and Edeka) are among the largest in Europe and the world (Lebensmittelzeitung, 1998).

In the late 1990s new methods of selling food gained importance. Examples are sales by electronic commerce which is expected to expand further in the next few years. A significant increase of food sales in new retail outlets like petrol stations, railway stations and airports can also be observed.

In most cases the large food retail chains took over transportation and logistics from the food industry, i.e. they supply their outlets with their own fleet of trucks. A reduction of transport costs, the delivery of full assortments for an outlet through only one carrier, and the direct control of the delivery chain up to the outlet are the main reasons for this development. In addition, merchandise information systems have been established in the outlets allowing retail management to have daily control of the turnover of all listed articles. As a consequence, the risk for producers that their items, especially new items, with low turnover rates will be removed from the assortment is now much greater. The pressure on the food industry resulting from organisational changes in the retail trade has even been intensified as some food chains recently introduced the Efficient Consumer Response (ECR) concept (Spannagel and Trommsdorf, 1999). It is expected that the relationship between the food industry and food retail trade will continue to change significantly in the years ahead.

Consequences of the Developments in the Food Retail Trade for Food Industry and Agriculture

The above mentioned developments in German food retail trade have severe impacts for the downstream subsectors of the food chain. Based on strong bargaining power, the food trade is more or less fixing purchase prices for the products delivered by the food industry. In addition the food retail trade forces suppliers to accept increasing price deductions, supports to the advertising expenses of the food trade, and specific payments for including the product into the assortment of the outlets. Furthermore, for the food retail trade, quality requirements have been extended and there is now demand for just-in-time supply.

As a consequence of the national and international concentration processes only large suppliers are adequate partners for these food retail companies. Purchasing activities of these food trade giants are continually centralised which means that large quantities are needed for the great number of outlets. Under these circumstances small- and medium-sized enterprises are attractive partners only if they offer interesting innovations, if they are able to establish -(at least regionally) close customer-supplier relationships, or if they establish effective strategic alliances to combine their output. Nevertheless, the need for structural changes in the food industry is still growing. The food industry's attempts to form strategic alliances have not yet been very successful.

The strengthening of the introduction of private labels by the food retail trade (their market share for food in Germany was 14.0% in 1999) consequently forced even leading suppliers of brands to produce private labels in spite of the risks associated with such activities. Due to higher profit margins it is expected that the percentage of private labels will further increase.

Marketing of Agricultural and Food Products

On the one hand marketing of agricultural and food products is performed by various actors in the food chain offering consumer-ready products, i.e. farmers, food handicraft enterprises, the food industry, and the food retail trade. On the other hand, cooperative group marketing and centralised sector marketing is of some importance.

The application of marketing instruments by farmers is limited to the selling of goods directly to final consumers. This is especially important for the direct sales discussed earlier. Marketing activities are therefore adjusted to this kind of local business. For example, promotion activities take place primarily at the point of sale. The marketing activities of the food handicraft enterprises are of a similar nature.

Most important are the marketing activities of the food industry employing all kinds of marketing measures. Taking into account the given framework of the food market, a high commitment to marketing is a necessary prerequisite for success and competitiveness. It has been already pointed out that marketing activities of the food industry (e.g. price policy measures) have to take into account the limits set by the powerful food trade. In the following, some information is presented on product innovation and advertising as an example.

According to annual surveys during the last ten years marketing managers of German industry judge product innovation as the most important prerequisite for the success of their company in the following year (Stippel, 1998: 106). In spite of this positive evaluation, in 1998 only 397 million DM were spent for research and development (R&D) in the German food industry, representing 0.2% of the annual turnover (Grenzmann and Wudke, 1999: 4). It is questionable whether this relatively small amount is sufficient to finance the expensive research necessary for products like functional foods and nutraceuticals for which growing demand is forecasted.

Recently the number of food innovations introduced to the German market shows a decreasing trend: 1,632 new products were introduced in 1996, 1,253 in 1997, and 849 in 1998 (Lebensmittel-Praxis, several years). New drinks, confectionery, and dairy products were the product areas with the largest numbers of product innovations. According to the same survey, around 50% of the new products introduced were withdrawn from the market in the same year because of a lock of market success.

In the given competitive market situation of the German food sector, advertising is one of the most important marketing measures to establish unique positioning for brands and to influence consumers' attitudes. Furthermore, if product innovations are introduced to the market, advertising and other promotional measures are necessary to attain sufficient recognition of the new product by consumers.

Therefore, it is reasonable that the German food industry ranks second in media spending in Germany. In absolute terms, advertising expenditures for food and drinks increased by 10.7% from 4.762 billion DM in 1994 to 5.271 billion DM in 1998 (AC Nielsen, 1999). In 1998 the producers of beer and beverages made the highest advertising expenditures with 1.336 billion DM, the makers of chocolate and sugar confectionery with 1.135 billion DM, and processors of dairy products with 491 million DM in. Relatively, the whole food sector is spending about 2.3% of its turnover for advertising. Significantly above the average are the advertising expenditures of the producers of chocolate and sugar confectionery (7.9%), beer (4.6%), and edible oils and fats (4%). The media selection for advertising food differs strongly from the media selection of other branches. In 1998 the nationwide market share of television advertising was 42.8%, less than the print media with 48.9%. In the food sector, television advertising dominates: the market share increased from 51.6% in 1988 to 84.7% in 1998. One explanation might be that advertising for food products aims mainly at communicating experience attributes (Becker and Buchardi, 1996: 50).

Because of the limited opportunities for marketing by individual farms as well as by small- and medium-sized food industry companies, a central marketing company (CMA) was founded according to the Marketing Fund Law of 1969. It is mainly funded by fixed obligatory charges per product unit which the processors have to pay. However, as a central marketing organisation, the CMA has to assume a neutral position in competitive matters and must balance the interests of the various participating economic groups (Besch, 1993: 27-28). Consequently, it can fulfil only supportive tasks as its activities are restricted to basic and export market research, basic and generic advertising, support of sales promotion activities, and exhibitions. In addition educational activities, extension, and informational programs on the different parties in the marketing channel can be carried out. An important activity of the CMA is the establishment of quality label programs (Gütezeichen- and Gütesiegelprogramme) supported by comprehensive and rigid quality controls.

In addition to the central marketing by the CMA, other cooperative marketing institutions were founded during recent decades (see Balling, 1997). Some of them were established by the Laender (e.g. the marketing unit of the Bavarian Ministry of Agriculture), others by interest groups for specific products (e.g. the German Wine Institute to promote German wine). The activities of these institutions are similar to the CMA, however they are limited to the promotion of specific products or origins.

Expenditures for cooperative marketing activities are estimated at 283 million DM for 1994 (Balling, 1997: 51). About 62% was spent by the CMA and 22% by product-specific organisations; the organisations for wine have the highest budgets in Germany. Total cooperative advertising expenditures for food in 1994 amounted to 199 million DM; of this amount 65 million was spent by the CMA, 7 million by regional German organisations and 137 million by foreign cooperative market institutions. Cooperative advertising expenditures decreased to 162 million DM in 1998; 39 million DM by CMA, 12 million DM by regional German organisations, and 101 million by foreign cooperative marketing institutions which reduced cooperative advertising activities on the German market during that period (BIK Aschpurwis + Behrens, 1999).

Challenges for the Future

For the successful development of the German food chain in the coming years, several challenges must be managed and solutions found. The following seem to be of particular importance:
- For the competitiveness of the value-added chain it is essential that all subsectors are themselves competitive. The competitiveness of the whole chain can be significantly improved if the vertical integration of the subsectors is intensified. To track the wishes and attitudes from the final consumers through the whole chain makes close co-ordination between the various partners necessary. Inte-

gration processes have to be intensified; for example, by applying the ECR concept to the food trade and the food industry, but also by vertically integrating the food industry and agriculture.
- Today the acceptance of food fundamentally depends on high quality and other positive features of the goods produced and offered to domestic and foreign consumers and the utility they incorporate for the consumer. First, it seems necessary that the focus on consumer wishes be strengthened and become the ultimate basis for decisions made by the actors in the food chain. To transform consumer wishes into creative products, the expenditures for R&D must be increased. Secondly, in light of various food scandals, it is essential that the consumer can be guaranteed safe, healthy food products. An important step has been taken in this direction by establishing quality management systems in many German food industry companies. However, it seems necessary that most farmers as well as food trade companies also establish such systems in the future.
- It is to be expected that aside from quality, price also be a key buying criteria in the future. Low prices in combination with sufficient profits for the actors in the food chain can only be realised with high efficiency of the vertically connected enterprises and with an economic framework to produce and distribute food as cheaply as possible. It has been mentioned that the professionalism of the participants in the food chain in Germany must be improved. Intensifying education, continuing the training of managers and employees, and a more extensive use of modern decision aids can assist this process. However, in addition to low-cost production and distribution, strong efforts are necessary to improve the size structure in agricultural production and the food industry. There are many further necessary steps to improve the competitiveness of the German value-added chain in the future. These include the realisation of economies of scale and scope (including pecuniary economies of scale) and of the advantages of large-scale enterprises in R&D and marketing, in addition to the exploitation of learning curve effects.

References

AC Nielsen (1999). Werbetrends in den klassischen Medien 1998. *Horizont.* Hamburg: A.C. Nielsen Werbeforschung S+P GmbH.

Balling, R. (1997). *Gemeinschaftsmarketing für Lebensmittel.* Kiel: Vauk.

Becker, T. and Buchardi, H. (1996). Möglichkeiten und Grenzen der Lebensmittelwerbung. Diskussionsbeitrag 9612. Göttingen: Institut für Agrarökonomie.

Besch, M. (1993). Agricultural Marketing in Germany. *Journal of International Food & Agribusiness Marketing.* 5 (3/4): 5-35.

Bik Aschpurwis + Behrens GmbH (1999). Werbeaufwendungen der CMA, der regionalen Werbegemeinschaften und ausländischer Anbieter auf Basis der Nielsen Werbeforschung S+P, Klassische Medien. Hamburg.

Bundesministerium für Ernährung, Landwirtschaft und Forsten (several years). *Agrarbericht der Bundesregierung.* Bonn: Universitäts-Buchdruckerei.

Bundesregierung (1998). Stärkung der Agrareinkommen durch eine verbesserte Marktposition. Antwort auf eine Große Anfrage der SPD-Fraktion. *Agra-Europe* 39 (3): 1-34.

Bundesvereinigung der Deutschen Ernährungsindustrie e.V. (1999). *Jahresbericht '98'99.* Bonn: Universitäts-Buchdruckerei.

CMA Mafo Briefe (1999). *Aktuelles aus der Agrarmarktforschung* 5: 1.

Deutscher Fleischer-Verband (1998). *Geschäftsbericht 1998.* Frankfurt am Main.

Deutscher Raiffeisenverband e.V. (1999). *Jahrbuch 1998.* Bonn.

Deutscher Raiffeisenverband e.V., Deutscher Bauernverband e. V. (1998). Gemeinsame Initiative des Deutschen Raiffeisenverbandes e.V. und des Deutschen Bauernverbandes e.V. zur Strukturanpassung der Genossenschaften. *Welt der Milch* 52 (6): 202-204.

Elsinger, M. (1991). Erzeugergemeinschaften als Organisationsmodell zur Förderung eines marktgerechten Agrarangebots. *Dissertation.* Freising-Weihenstephan.

Ernährungsdienst (1999). *Der Agrarmarkt in Deutschland 1999.* Frankfurt am Main: Strothe.

Grenzmann, Ch. and Wudke, J. (1999). Wirtschaft investiert wieder kräftig in FuE. Stifterverband für die Deutsche Wissenschaft (eds): *FuE Info 1*: 2-8.

Halk, O., Franken, R., Gödeke, K. and Dwehus, J. (1999). Erfolgsfaktoren von Erzeugergemeinschaften - Ergebnisse einer empirischen Untersuchung. In: Landwirtschaftliche Rentenbank (eds*): Innovative Konzepte für das Marketing von Agrarprodukten und Nahrungsmitteln.* Frankfurt am Main: 51-92.

Hartmann, M. (1993). Überlegungen zur Wettbewerbsfähigkeit des Deutschen Ernährungsgewerbes. *Agrarwirtschaft* 42: 237-247.

Herrmann, R., Reinhardt, A. and Zahn, C. (1996). Wie beeinflußt die Marktstruktur das Marktergebnis. *Agrarwirtschaft* 45: 186-196.

Kallfass, H. H. (1993). Kostenvorteile durch vertikale Integration im Agrarsektor. *Agrarwirtschaft* 42: 228-237.

Kuhnert, H. (1998). *Direktvermarktung in konventionell und ökologisch wirtschaftenden Betrieben.* Kiel: Vauk.

Lebensmittel-Praxis (several years). Die erfolgreichsten Neuen im Handel. *Lebensmittel-Praxis.*

Lebensmittelzeitung (1998). Big Player nutzen ihren „Heimvorteil". *Lebensmittelzeitung* 50 (34): 4.

Lebensmittelzeitung (1999a). Die größten Lieferanten des Handels. *Lebensmittelzeitung* 51 (33): 10

Lebensmittelzeitung (1999b). Die größten Handelsunternehmen der Branche 1998. *Lebensmittelzeitung* 51 (9): 4.

Lingenfelder, M. and Lauer, A. (1999). Die Unternehmenspolitik im deutschen Einzelhandel zwischen Währungsreform und Währungsunion. In: Dichtl, E.and Lingenfelder, E. *(eds), Meilensteine im Deutschen Handel.* Frankfurt am Main: Deutscher Fachverlag: 11-55.

Nestlé Deutschland AG (1999). *Ernährung in Deutschland: Gesund essen – gesund leben. Nestlé Studie zur Anuga 1999.* Frankfurt am Main: Nestlé.

Nienhaus, A. (1995). On the change in consumer behavior – New insights, background and opportunities. IDF (eds): *Milk Policy on Trial.* Materials of an IDF Symposion. Vienna, 12-13 September 1995.

Ottowitz, T. (1997). *Qualitätsmanagment bei der Vermarktung von Fleisch.* Kiel: Vauk.

Spannagel, R. and Trommsdorf, V. (1999). Kundenorientierung im Handel. In: E. Dichtl, E. and Lingenfelder E. *(eds). Meilensteine im Deutschen Handel.* Frankfurt am Main: Deutscher Fachverlag: 57-88.

Statistisches Bundesamt (1998). *Statistisches Jahrbuch für die Bundesrepublik Deutschland.* Frankfurt am Main: Metzler-Poeschel.

Statistisches Bundesamt (several years). *Statistisches Jahrbuch über Ernährung, Landwirtschaft und Forsten der Bundesrepublik Deutschland.* Münster-Hiltrup: Landwirtschaftsverlag.

Stippel, P. (1998). Top-Thema: Was bringt 1999? *Absatzwirtschaft* 9: 106.

Weindlmaier, H. (1994). Entwicklung der deutschen Molkereigenossenschaften - Problem-analyse und Zukunftsperspektiven. *Deutsche Milchwirtschaft* 45 (14): 632-637; 45 (15): 703-707.

Weindlmaier, H. (2000). Die Wettbewerbsfähigkeit der deutschen Ernährungsindustrie: Methodische Ansatzpunkte zur Messung und empirische Ergebnisse. In: v. Alvensleben, R., Koester, U. and Langbehn, C. (eds): Wettbewerbsfähigkeit und Unternehmertum in der Land- und Ernährungswirtschaft. Schriften der Gesellschaft für Wirtschafts- und Sozialwissenschaften des Landbaues e.V., Band 36. Münster-Hiltrup: Landwirtschaftsverlag.

Weindlmaier, H., Kochan, A. and Petersen, B. (1997). Notwendigkeit von Qualitätsmanagement in der deutschen Ernährungswirtschaft. In: Forschungsgemeinschaft Qualitätssicherung e.V. (eds). *Einführung von Qualitätsmanagementsystemen nach ISO 9000 ff. in der landwirtschaftlichen Produktion und im Nahrungs- und Genußmittelgewerbe.* Berlin: Beuth.

Wendt, H. (1991). Anpassungen in der Ernährungswirtschaft an Änderungen im Lebensmittelrecht. In: Schmitz, P. M. and Weindlmaier, H. (eds). Land- und Ernährungswirtschaft im europäischen Binnenmarkt und in der internationalen Arbeitsteilung. Schriften der Gesellschaft für Wirtschafts- und Sozialwissenschaften des Landbaues e.V., Band 21. Münster-Hiltrup: Landwirtschaftsverlag: 209-217.

Wirthgen, B. and Maurer, O. (1992). Direktvermarktung. Stuttgart: Ulmer.

Wöhlken, E. (1991). *Einführung in die landwirtschaftliche Marktlehre.* 3. Auflage. Stuttgart: UTB.

Zentralverband des Deutschen Bäckerhandwerks (1998). URL: http:/www.baeckerhandwerk.de.